# Participatory Action Research Approaches and Methods

Participatory Action Research (PAR) approaches and methods have seen an explosion of recent interest in the social and environmental sciences. PAR involves collaborative research, education and action oriented towards social change, representing a major epistemological challenge to mainstream research traditions. It has recently been the subject of heated critique and debate, and rapid theoretical and methodological development.

This book captures these developments, exploring the justification, theorisation, practice and implications of PAR. It offers a critical introduction to understanding and working with PAR in different social, spatial and institutional contexts. The authors engage with PAR's radical potential, while maintaining a critical awareness of its challenges and dangers. Part I explores the intellectual, ethical and pragmatic contexts of PAR; the development and diversity of approaches to PAR; recent poststructuralist perspectives on PAR as a form of power; the ethics of participation; and issues of safety and well-being. Part II is a critical exploration of the politics, places and practices of PAR, exploring methods including diagramming, cartographies, art, theatre, photovoice, video and geographical information systems are also discussed. Part III reflects on how effective PAR is, including the analysis of its products and processes, participatory learning, representation and dissemination, institutional benefits and challenges, and working between research, action, activism and change.

The authors find that a spatial perspective and an attention to scale offer helpful means of negotiating the potentials and paradoxes of PAR. The book adds significant weight to the recent critical reappraisal of PAR, suggesting why, when, where and how we might take forward PAR's commitment to enabling collaborative social transformation. It will be particularly useful to researchers and students of Human Geography, Development Studies and Sociology.

**Sara Kindon** is a Senior Lecturer in Human Geography and Development Studies at Victoria University of Wellington, Aotearoa New Zealand.

**Rachel Pain** is a social geographer at Durham University in the UK.

**Mike Kesby** is a Lecturer in Human Geography at the University of St Andrews, Scotland, UK.

## Routledge Studies in Human Geography

This series provides a forum for innovative, vibrant, and critical debate within Human Geography. Titles will reflect the wealth of research which is taking place in this diverse and ever-expanding field. Contributions will be drawn from the main sub-disciplines and from innovative areas of work which have no particular sub-disciplinary allegiances.

*Published:*

1. **A Geography of Islands**
   Small island insularity
   *Stephen A. Royle*

2. **Citizenships, Contingency and the Countryside**
   Rights, culture, land and the environment
   *Gavin Parker*

3. **The Differentiated Countryside**
   *Jonathan Murdoch, Philip Lowe, Neil Ward and Terry Marsden*

4. **The Human Geography of East Central Europe**
   *David Turnock*

5. **Imagined Regional Communities**
   Integration and sovereignty in the Global South
   *James D Sidaway*

6. **Mapping Modernities**
   Geographies of Central and Eastern Europe 1920–2000
   *Alan Dingsdale*

7. **Rural Poverty**
   Marginalisation and exclusion in Britain and the United States
   *Paul Milbourne*

8. **Poverty and the Third Way**
   *Colin C. Williams and Jan Windebank*

9. **Ageing and Place**
   *Edited by Gavin J. Andrews and David R. Phillips*

10. **Geographies of Commodity Chains**
    *Edited by Alex Hughes and Suzanne Reimer*

11. **Queering Tourism**
    Paradoxical performances at gay pride parades
    *Lynda T. Johnston*

12. **Cross-Continental Food Chains**
    *Edited by Niels Fold and Bill Pritchard*

**13. Private Cities**
*Edited by Georg Glasze, Chris Webster and Klaus Frantz*

**14. Global Geographies of Post Socialist Transition**
*Tassilo Herrschel*

**15. Urban Development in Post-Reform China**
*Fulong Wu, Jiang Xu and Anthony Gar-On Yeh*

**16. Rural Governance**
International perspectives
*Edited by Lynda Cheshire, Vaughan Higgins and Geoffrey Lawrence*

**17. Global Perspectives on Rural Childhood and Youth**
Young rural lives
*Edited by Ruth Panelli, Samantha Punch, and Elsbeth Robson*

**18. World City Syndrome**
Neoliberalism and inequality in Cape Town
*David A. McDonald*

**19. Exploring Post Development**
*Aram Ziai*

**20. Family Farms**
*Harold Brookfield and Helen Parsons*

**21. China on the Move**
Migration, the state, and the household
*C. Cindy Fan*

**22. Participatory Action Research Approaches and Methods**
Connecting people, participation and place
*Sara Kindon, Rachel Pain and Mike Kesby*

*Not yet published:*

**23. Time Space Compression**
Historical geographies
*Barney Warf*

**24. International Migration and Knowledge**
*Allan Williams and Vladimir Balaz*

**25. Design Economies and the Changing World Economy**
Innovation, production and competitiveness
*John Bryson and Grete Rustin*

**26. Whose Urban Renaissance?**
An international comparison of urban regeneration policies
*Libby Porter and Katie Shaw*

**27. Tourism Geography**
A new synthesis, second edition
*Stephen Williams*

*Frontispiece* I am here. Self Portrait is about who you are. Who you are is what you believe you are. It's me, Gaby Kitoko. I'm here ready to act. Never doubt – we can make a difference together (Credit: Gaby Kitoko for Media 19/ Self Portrait Refugee).

# Participatory Action Research Approaches and Methods

## Connecting people, participation and place

Edited by
Sara Kindon, Rachel Pain
and Mike Kesby

LONDON AND NEW YORK

Transferred to digital printing 2010

First published 2007 by Routledge
2 Park Square, Milton Park, Abingdon, Oxon OX14 4RN

Simultaneously published in the USA and Canada
by Routledge
270 Madison Avenue, New York, NY 10016

*Routledge is an imprint of the Taylor & Francis Group, an informa business*



Typeset in Times New Roman by
Bookcraft Ltd, Stroud, Gloucestershire



*British Library Cataloguing in Publication Data*
A catalogue record for this book is available from the British Library

*Library of Congress Cataloging in Publication Data*
Participatory action research approaches and methods : connecting people, participation, and place / [edited] by Sara Kindon, Rachel Pain and Mike Kesby.
p. cm.
Includes bibliographical references and index.
1. Social participation—Research—Methodology. 2. Political participation—Research—Methodology. 3. Communities—Research—Methodology. I. Kindon, Sara Louise. II. Pain, Rachel. III. Kesby, Mike, 1966–
HM711.P37 2007
302'.1401—dc22 2007024739

ISBN10: 0-415-40550-5 (hbk)
ISBN10: 0-415-59976-8 (pbk)
ISBN10: 0-203-93367-2 (ebk)
ISBN13: 978-0-415-40550-8 (hbk)
ISBN13: 978-0-415-59976-4 (pbk)
ISBN13: 978-0-203-93367-1 (ebk)

**We dedicate this book to Janet Townsend to honour her work in pioneering feminist and participatory approaches in Geography, and to acknowledge the profound impacts of her critical intellect and gentle manner on our own thinking and practice.**

# Contents

# Boxes

# Figures

# Plates

# Contributors

**Mags Adams**, Senior Research Fellow, University of Salford, UK is particularly interested in theoretical interconnections between sustainability, urban form and individual practice. Mags works with innovative methodologies including photo-surveys and soundwalks to highlight sensory experience and to develop research around sensory urbanism.

**Catherine Alexander** is a Research Associate and PhD student at the Department of Geography, Durham University, UK. Her research is concerned with issues of fear, crime, intergenerationality, and the intersections between social identities and social exclusions.

**Natalie Beale** is a PhD student at the Department of Geography, Durham University, UK. She is researching young people's understandings of health and health risks.

**Nathan Brightbill** is an associate with the Pomegranate Centre, a non-profit organisation that focuses on community participation to construct gathering places and public art. He holds an MSc in Community Development from UC Davis, and is an MLA candidate at the University of Washington, USA.

**Caitlin Cahill** is committed to engaged interdisciplinary scholarship. She is an Assistant Professor of Community Studies at the University of Utah State, USA. She received her PhD in Environmental Psychology with a concentration in public policy and urban studies from the City University of New York.

**Jenny Cameron** works in Griffith University's School of Environment. She has conducted PAR in Australia, working with academic colleagues, staff from institutions and community members, with results communicated through academic publications, a resource kit and documentary (see www.communityeconomics.org).

**Paul Chatterton** is a Lecturer in Human Geography at the University of Leeds, UK, where he coordinates an MA programme in 'Activism and Social Change'. His research interests include urban culture, protest and social movements and sustainable and international development.

**Marie Cieri** is Assistant Professor of Geography at The Ohio State University and former director of The Arts Company of Cambridge, MA, USA. She has extensive experience as a producer, curator, consultant and writer in the performing, visual and media arts.

**Chris Coe** is a Senior Research Fellow in the Institute of Health, University of Warwick, UK. Her research interests lie in the area of health inequalities, in particular child and family heath and also in participatory research methods.

**Catherine Dolan** is an anthropologist lecturing at the Said Business School, University of Oxford. Her research focuses on how global shifts in consumer preferences, product standards and new structures of governance impact on production methods, labour and gender relations in Africa.

**Sarah Elwood** is an Assistant Professor of Geography at the University of Washington, USA. Her research intersects urban geography, GIScience, and qualitative and participatory research methods; with a particular focus on the social and political impacts of geospatial data and technologies.

**Ruben Feliciano** is the Director of Community Development at the Near Northwest Neighborhood Network, Chicago, IL, USA.

**Duncan Fuller** is Principal Lecturer in Human Geography and University Enterprise Fellow at Northumbria University, Newcastle upon Tyne, UK. His research interests focus around emerging economic geographies of social and financial inclusion/exclusion, credit unions, academic activism and participatory methodologies.

**Michael Gavin** teaches in the Environmental Studies Programme at Victoria University of Wellington, New Zealand, and conducts research on integrated conservation and development issues in Amazonia and the Pacific.

**Kathleen Gems** is the Development Director at the Near Northwest Neighborhood Network, Chicago, IL, USA.

**Nandhini Gulasingam** is the Technology Coordinator at DePaul University's Monsignor John J. Egan Urban Centre, Chicago, IL, USA.

**Fungisai Gwanzura-Ottemoller** graduated from St Andrews University, Scotland, with a PhD in Geography in 2005 having completed a thesis on children's sexual knowledge and behaviour in Zimbabwe. She currently works for the charity Children in Scotland as a research officer.

**Thomas Haenga Curtis** affiliates to the Māori tribes of Te Arawa, Tainui, Ngāti Rangitihi, Ngāti Hauiti and Tuwharetoa in Aotearoa/New Zealand. He has been working in the IT industry for twenty-five years. He is also a part-time radio journalist and amateur videographer.

**Madeleine Hall-Arber** is an Anthropologist with a research and outreach focus on the mitigation of negative impacts of regulatory changes in management on communities and the commercial fishing industry in the USA.

**Nancy Hansen** has a PhD from the Department of Geographical and Earth Sciences, University of Glasgow, Scotland. Her work examines employment issues and disabled women.

**Jane Higgins** is a Senior Research Fellow at Lincoln University, New Zealand. Her research explores the experiences of young people in transition between school and post-school worlds, youth (un)employment and the dynamics of the youth labour market.

**The Holy Family Research Team** wish to remain anonymous, but consists principally of two women in their twenties from 'Holy Family Settlement', Fiji.

**William Howard** is the Executive Director of the West Humboldt Park Family and Community Development Council, Chicago, IL, USA.

**Geoffrey Hume-Cook** is an Australian screen producer with a background in cultural studies, based in Wellington, Aotearoa/New Zealand. He is consciously engaging with his *manuhiri* status (visitor), through ongoing work with Ngāti Hauiti and individuals from other *iwi* (Māori tribes).

**James Ndwiga Kathuri** is a Geography Lecturer in the Department of Education and Counselling at Kenya Methodist University, Meru, Kenya. His research interests include climate change and health, gender, horticultural industry, and participatory methodologies.

**Mike Kesby** is a Human Geographer at St Andrews University, Scotland. He is conducting Participatory Action Research with young people in Zimbabwe on questions of sexual health and HIV.

**Sara Kindon** lectures in Human Geography and Development Studies at Victoria University of Wellington, Aotearoa/New Zealand. She collaborates with indigenous Māori, refugee-background communities and young people using Participatory Action Research. Her research interests focus on issues of visuality, complicity and desire.

**Brigette Krieg** is a *Metis* woman from Prince Albert, Saskatchewan, Canada. She is currently a PhD candidate with the University of Calgary, Canada and a Faculty/ Programme Coordinator with First Nations University of Canada. Her interest areas are anti-oppressive practice, social justice, marginalisation, Indigenous issues, and women's issues.

**Robbie McCauley** won an OBIE Award for *Sally's Rape*, acted in Ntozake Shange's *For Colored Girls Who Have Considered Suicide When the Rainbow Is Enuf*, directed Daniel Alexander Jones' *Belle Canto* premiere. She is Associate Professor of Performing Arts at Emerson College, Boston, MA, USA.

**Hazel McFarlane** is a Research Fellow within the Strathclyde Centre for Disability Research, University of Glasgow, Scotland. Her research interests include disability history, disabled women's access to reproductive choices and access to health related services.

**Reid Mackin** is the Director of Industrial and Workforce Development at the Greater Northwest Chicago Industrial Corporation, Chicago, IL, USA.

**Julia McMillan** is a PhD student at the Department of Geography, Durham University, UK. She is researching children's experiences and views of the journey to school and urban sustainability.

**Lynne Manzo** is an Environmental Psychologist and an Assistant Professor in the College of Architecture and Urban Planning at the University of Washington, Seattle, USA. Her research focuses on place meaning, identity and the politics of place.

**Eliud Medina** is the Executive Director of the Near Northwest Neighborhood Network, Chicago, IL, USA.

**Gemma Moore** is a Research Assistant at The Bartlett School of Graduate Studies, University College London, in the UK. Her research interests include urban sustainability, environmental quality and community involvement. She is currently undertaking a PhD.

**Karen Nairn** is a Senior Lecturer in Education at the University of Otago, New Zealand. Her research explores processes of exclusion in education shaped by gender, sexuality and race, and young people in transition from school to post-school lives.

**Ruth Newport** is completing a Masters of Communication at Auckland Institute of Technology, Aotearoa/New Zealand.

**Maggie Opondo** is a Lecturer in the Department of Geography and Environmental Studies, University of Nairobi, Nairobi, Kenya. Her research interests include gender and labour rights in global supply chains, ethical trade and corporate social responsibility and adaptability and vulnerability to climate change.

**Rachel Pain** is Social Geographer at Durham University in the UK. Her research interests lie around fear, well-being and social justice. She is currently conducting Participatory Action Research with young asylum seekers, refugee and locally-born young people in North East England.

**The Philippine Women Centre of BC and *Ugnayan ng Kabataang Pilipino sa Canada***, founded in 1989 and 1995 respectively, are grassroots organisations dedicated to the empowerment and genuine development of the Filipino community through research, education, organisation and mobilisation.

**Joyce Potaka** has lived and worked in Aotearoa/New Zealand and Australia. She is the mother of five children and grandmother of four. One of her greatest dreams is to see her children and grandchildren reconnected with their *whenua* (land).

**Geraldine Pratt** is Professor in Geography at the University of British Columbia, Vancouver, BC, Canada. Her most recent books are *Working Feminism* (2004) and *The Global and the Intimate* (edited with Victoria Rosner, 2006). She co-edits the journal *Society and Space*, as well as *The Dictionary of Human Geography* (4th and 5th editions).

**Niuris Ramos** is the Lead Community Organiser at the Near Northwest Neighborhood Network, Chicago, IL, USA.

**Lana Roberts** is a First Nations *Cree* and *Metis* woman from northern Saskatchewan. She is a Bachelor of Social Work student, a crisis worker and a photographer with the Prince Albert Photovoice group. She is interested in working to help improve living conditions in all aspects of First Nations people.

**Paul Routledge** is a Reader in Human Geography at the Department of Geographical and Earth Sciences at the University of Glasgow. His research interests include global justice networks; resistance movements; and geopolitics. He is co-editor of *The Geopolitics Reader*.

**Kevin St. Martin** is a Geographer interested in critical analyses of economic and resource management discourse. He uses Geographic Information Systems (GIS) and participatory methods to foster alternative and community-centred approaches to development and resource management.

**Eleanor Sanderson** has recently completed a PhD in Geography from Victoria University of Wellington, Aotearoa/New Zealand. She has a background in development and currently works both within the development NGO sector and as a priest in the Anglican Church.

**Sobeida Sierra** is the Office Manager of the West Humboldt Park Family and Community Development Council, Chicago, IL, USA.

**Judith Sligo** is a Research Assistant at the University of Otago, New Zealand. Her work, on two long-term projects, explores young people's identities at the end of their compulsory schooling, and parents' experiences of parenting pre-school children.

**Maria Stuttaford** is an Honorary Lecturer in Geography, University of St Andrews and Associate Research Fellow, Institute of Health, University of Warwick, both in the UK. Her research and teaching are mainly in the areas of health and human rights and participatory research methods.

**Adrian Tangaroa Wagner** co-directs the Wâhû Creations entertainment business, Wellington, Aotearoa/New Zealand. He is also an announcer for Te Upoko o Te Ika 1161 am contributing 'to the survival of the Māori language and to bringing to the people, issues affecting Māori today'.

**Divya Tolia-Kelly** lectures at Durham University, UK. Her research interests include visual cultures of landscape, material cultures, memory, heritage and national identity. She has been collaborating with artists since 1997, resulting in several art exhibitions in the UK and USA.

**María Elena Torre** is the Chair of Education Studies at Eugene Lang College, The New School, New York, US. Committed to Participatory Action Research in schools, prisons and communities, she also consults with college, government, and community groups interested in establishing college-in-prison programmes.

***Umaki* research participants** wish to remain anonymous, but are members of Mothers' Union groups in Tanzania.

**Miguel Vasquez** works with communities in the buffer zone of the Cordillera Azul National Park as an extension agent for the Peruvian conservation organisation CIMA-Cordillera Azul.

**Alaka Wali** is the John Nuveen Curator in Anthropology and Director of the Center for Cultural Understanding and Change (CCUC) at the Field Museum in Chicago, IL, USA.

**Senorina Wendoh** taught postcolonial literature at the University of Nairobi, Nairobi, Kenya, before moving to the development sector. She has researched among and worked with local NGOs and grassroots communities on African perspectives on gender. Senorina is currently a freelance consultant working on north–south partnerships.

**Kirsty Woods** works in resource management policy. She has worked on fisheries policy with Te Ohu Kaimoana, the Māori Fisheries Trust, Aotearoa/New Zealand. She is currently taking leave to study photography, particularly to develop her skills in documentary photography.

**Friederike Ziegler** is a PhD student at the Department of Geography, Durham University. Her research focuses on old age, mobility and social exclusion.

# Foreword

In this era of neoliberal politics and funding, many academics feel pressured by the ambitions and needs of their universities to prioritise rapid publication of theoretical research in high status academic journals. It is thus salutary to see the work of social and environmental researchers represented in this volume, which has as its motivation the desire to foster social justice and change by working collaboratively with communities. The contributors are committed to disseminating their work in multiple and non-traditional ways, in actions as well as in print. This is not to say that they fail to make theoretical and conceptual advances. Indeed, as the book demonstrates, the approaches and insights of Participatory Action Research challenge ways of thinking, learning and being in the world, both among researchers and those members of communities with whom they collaborate.

Participatory Action Research espouses large goals, not easily achievable. Working across the boundaries of academia and other worlds requires cultivation of mutual understanding and respect, sensitivity to differences in organisational cultures and goals, networking and sharing information, recognising and strengthening individual and group capacities, questioning priorities, formulating questions so as to foster change and not simply to'explain' what is, and, not surprisingly, dealing with diverse personalities. Not least of the challenges are to identify necessary financial and other resources and to negotiate how these will be shared and managed in transparent ways. What degrees of freedom do partners have? Can actions be scaled up from local settings to larger arenas, and if so, how? Participatory Action Research is not an approach that can be rushed into, but one that takes time and talent, that requires the building of trust, and being sensitive to 'turf'.

In bringing together an array of authors who have wide experience of engaging in Participatory Action Research in diverse settings, Sara Kindon, Rachel Pain and Mike Kesby offer a guide which, as they write in their introduction, presents a sense of the approach's 'radical potential while maintaining a critical awareness of its challenges and dangers'. The chapters address a wide range of situations, consider ethics and politics, both personal and institutional, illustrate numerous methods, many of which involve visual approaches, and highlight the significance of space, place and scale in fostering social and environmental change. Contributors include community collaborators writing with the academic researchers, practising what they preach, presenting their experiences and reflections in readily accessible prose. Throughout we are offered not only answers, but are also prompted to maintain flexibility, to adapt, to reflect and rethink, yet to persist in seeking ways to work together to engage in research actions that will advance social change and justice.

Professor Janice Monk
University of Arizona
Tuscon, USA

# Acknowledgements

This book was originally the concept of one of its editors, Sara Kindon, but quickly became a collective enterprise as Rachel Pain and Mike Kesby came on board. We worked together for the first time on a book chapter discussing participatory approaches for a textbook on research methods in Human Geography. We found the experience of writing together to be both fun and productive; it laid the foundation for this volume. It has been a joy working on this book together and we look forward to future collaborations both in, and outside of, print.

We are grateful to the contributors for their commitment and patience. Their innovative work and concise reflections on it have inspired our own writings and practice. It has been exciting to develop a network with experienced and new researchers through the project of this book, and by extension, with the co-researchers and communities with whom they work.

We want to thank Andrew Mould at Routledge for supporting our vision for this book and Jennifer Page for her gentle reminders along the path to its completion. We appreciate the grant from the Faculty of Science, Architecture and Design at Victoria University of Wellington, which funded the editorial work. Most especially we wish to thank Kate Satterthwaite for all of her excellent work. She took on the formidable task of editing and ensuring consistency with a graceful dedication and keen eye.

We are grateful to Sage Publications Ltd for permission to reproduce in Chapter 22 sections from the article: 'The "learning" component of Participatory Learning and Action (PLA) in health research: reflections from a local SureStart evaluation', *Qualitative Health Research* (forthcoming).

We are also grateful to the following for kind permission to reproduce their photographs: Gaby Kitoko and Media19 for the front plate; the Ethical Trading Initiative 'ETI smallholder guidelines 2005' for Plate 10.1; Melanie Carvalho for Plates 16.1 and 16.3; the Fed Up Honeys for Figure 23.1.

We each have some personal acknowledgements as well.

*Sara*: I would like to extend a heartfelt thank you to Geoff Hume-Cook, for his visible and invisible contributions to this book as co-researcher, co-author, life partner and co-parent. I would like to thank our son, Mātai, for his zest for life and for introducing me to the joys of trolley buses. The contributions of many students and colleagues to my thinking over the years has been valuable, particularly students in my graduate paper on Young People and Participatory Development and the dear members of my academic women's writing group. Finally, to my whānau in Ngāti Hauiti, I offer a warm '*Kia ora*'.

*Rachel*: To the co-researchers, friends, students and colleagues I am lucky to know who have enriched this book – thank you. Love, as always, to Rob, Ben and Anna.

*Mike*: I would like to thank Rachel and Sara for their energy and seemingly boundless patience in the face of my stuttering and always delayed efforts; all the people in Zimbabwe who have so inspired my thinking over the years and for whom the present is such a terrible struggle; and my own family, Dale, Dan and Tim for supporting me through the tough times and for constantly reminding me that I should work to live, not …

# 1 Introduction

## Connecting people, participation and place

*Sara Kindon, Rachel Pain and Mike Kesby*

### Purpose and scope

Participatory Action Research (PAR) is an umbrella term covering a variety of participatory approaches to action-oriented research. Defined most simply, PAR involves researchers and participants working together to examine a problematic situation or action to change it for the better (Wadsworth 1998). For over seventy years, advocates of participatory approaches have been challenging the traditionally hierarchical relationships between research and action, and between researchers and 'researched' (Wadsworth 1998). They have sought to replace an 'extractive', imperial model of social research with one in which the benefits of research accrue more directly to the communities involved. Put another way, advocates have attempted to remove hierarchical role specifications and empower 'ordinary people' in and through research. Their intention is to transform an alienating 'Fordist' mode of academic production into a more flexible and socially owned process.

For a long time, this struggle occurred at the margins of the academy, but over the last decade or so a series of shifts in philosophical critique, economic policy and international geopolitics has generated a context in which participation can 'come in from the cold' (see Cornwall and Pratt 2003; Fals-Borda 2006a; Hall 2005). Furthermore, as millennial reflection caused researchers once again to question their role and relevance in a rapidly changing world (see Staeheli and Mitchell 2005), the academy has become more receptive toward a 'participatory turn' (Fuller and Kitchin 2004). Participatory and Action Research are rapidly becoming a leading paradigm within the social and environmental sciences (Brydon-Miller *et al.* 2004; Greenwood and Levin 1998; Jason *et al.* 2004; Park *et al.* 1993; Reason and Bradbury 2006; Selener 1997; Taggart 1997). The purpose of this book is to support researchers working within this paradigm.

As contributions to this book illustrate, the process of PAR is cyclical. Researchers and participants identify an issue or situation in need of change; they then initiate research that draws on capabilities and assets to precipitate relevant action. Both researchers and participants reflect on, and learn from, this action and proceed to a new cycle of research/action/reflection (see Kindon *et al.*, Chapter 2 in this volume). Together they develop context-specific methods to facilitate these cycles. These may include the adaptation of traditional social science methods like semi-structured interviews, focus groups and Geographic Information Systems (GIS), or innovations in visual or performative methods like diagramming, video and theatre (see Part II of this volume). This methodological openness reflects PAR's commitment to genuinely democratic and non-coercive research with and for, rather than on, participants (Pratt 2000; Wadsworth 1998).

Having said this, participatory approaches and PAR are not without their challenges and critics (Cooke and Kothari 2001a; Greenwood 2002; Hayward *et al.* 2004). Often, for academics to undertake participatory and action-oriented research, and to achieve a successful career, they must bridge 'two conflicting social worlds' (see Cancian 1993: 92). For example, the academy in many places continues to exclude the epistemologies involved in PAR (Kuokkanen 2004) and communities frequently question the relevance of academics to meet their needs (Stoeker 1999). Achievements often depend on researchers' commitment, creativity and imagination in negotiating competing discourses and expectations.

Criticism of participatory approaches, meanwhile, has intensified. For some, the increasing popularity of participation within development and policy contexts, for example, represents its commodification within schemes and research that remain 'top down' and extractive (see Cornwall and Brock 2005; Mohan 1999; Pain and Francis 2003; see also Kesby *et al.*, Chapter 3 in this volume). There are also considerable concerns about the under-theorisation of power, and the possibilities for marginalisation that occur within participatory processes striving for consensus and collective action (Cooke and Kothari 2001a).

While we are unapologetic advocates of participation, we believe (in line with PAR's long tradition of self-reflection and internal critique) that it is important to keep a critical eye open for its weaknesses, limitations and dangers. Thus we value contemporary academic critiques: we recognise that while participatory approaches seek socially and environmentally just processes and outcomes, they nevertheless constitute a form of power and can reproduce the very inequalities they seek to challenge (Cooke and Kothari 2001a; Kesby 2005). We acknowledge that the ubiquity of participation in international development, for instance, can make it seem like a tyrannical yet bland orthodoxy (Cooke and Kothari 2001a; Kapoor 2005). Further, we agree that participation has too often been dislocated from a radical politics oriented toward securing citizenship rights and informing underlying processes of social change (Hickey and Mohan 2005; see also Chatterton *et al.*, Chapter 25 in this volume). However, rather than abandon participation, our response is to look for ways in which such critique can fortify and transform our practice.

Indeed, Participatory Action Research is one means of repoliticising participation (Fals-Borda 2006a; Kapoor 2005). PAR emphasises dialogic engagement with co-researchers, and the development and implementation of context appropriate strategies oriented towards empowerment and transformation at a variety of scales. This political commitment has been posited as something of an antidote to the increasingly commonplace technocratic deployment of participation or participatory techniques (Kapoor 2005). Thus, while not a panacea for all research and development ills, there is much radical potential left in PAR, as the chapters in this volume illustrate. Our aim in this book, therefore, is to engage with PAR's radical potential, while maintaining a critical awareness of its challenges and dangers. For us, the strength of critically-informed PAR lies precisely in its ability to facilitate the intersections of theory, practice and politics between participants and researchers in a diversity of contexts.

As geographers, we find that a spatial perspective and an attention to scale offer helpful means of negotiating the potentials and paradoxes of PAR (Pain and Kindon 2007). The importance of space to social life is increasingly recognised across the social sciences (see Massey 2005) and the participatory development literature (Cornwall 2002; 2004b; Gaventa 2004; Kesby 2005; 2007a). The chapters in this volume remind us that space and place are important to participation as a political practice. They illustrate how understanding the spatialities of participation can inform both our theoretical understandings and the outcomes of social or environmental change.

Space is also important when trying to affect change beyond the various sites and arenas of participatory intervention (Cornwall 2002; Jones and SPEECH 2001; Kesby 2005; 2007a). Typically, participatory approaches prioritise local community concerns, the immediate social and natural environments in which they are located, and ground up processes. With greater attention to space and scale however, the local is understood as intimately connected to the global, regional, national, household and personal. PAR can help to unpick the hierarchical scaling of events, things and processes, conceptually, practically and politically (see Klodawsky 2007; Marston *et al.* 2005). It can help participants to re-engage with wider structures and processes of inequality to effect change. It can also involve and alter spaces of empowerment and action, when it contributes to policy, social or personal transformation (see also Kesby *et al.*, Chapter 3; and Pain *et al.*, Chapter 4 in this volume).

While the practice and discussion of PAR is by now widespread and well developed, there are few single texts to which readers can turn for information on every aspect of the approach: from underlying philosophy and ethics, through field techniques and practical guidance, to issues of dissemination, action outputs, activism and theoretical critiques of the approach itself (cf. Selener 1997). It is our purpose with this volume to provide such a text. We therefore take a wide-ranging and critical approach to consider the intersections of theory, practice and politics informing co-research using PAR. We recognise that PAR is a form of power, and return frequently to why we might nevertheless use participatory approaches and methods to address particular questions. We also consider how they might inform our research relationships and any resultant action. At the heart of these considerations, we attend to the practice of the ethical and spatial relationships involved, as they ultimately connect people, participation and place to the wider politics of social and environmental transformation.

We have therefore aimed to provide a book which:

- discusses ethical, personal and institutional challenges commonly encountered when using participatory approaches and methods;

- provides links between theoretically informed critiques of Participatory Action Research and its applied practice; and
- grounds the debates and practice within social and environmental sciences, and in particular foregrounds insights from Human Geography in exploring the value of a spatial perspective to the practice of participation.

## Format

The book is organised into three parts, each containing short, readable chapters. We want to disrupt the common perception that academic work is verbose, impenetrable and of little use to anyone in the 'real world', by speaking to the busy practitioner, the time-pressured graduate student and the community researcher. At the same time, we want to acknowledge and nurture the value of theory and academic reflection. Indeed, we know that many practitioners of PAR, both inside and outside the academy, are interested in keeping abreast of emerging theories and methodological developments in order to strengthen their disciplinary knowledge and applied practice. The key to a robust future for PAR is clear, respectful communication that closes the perceived gap between theorists and practitioners and further facilitates the informed use of PAR within, and beyond, the academy.

It will be obvious to readers that the inclusion of many images and figures is an attempt to reflect the importance and utility of visual methods within PAR. They also illustrate the people and places involved and some of the products produced. We hope that they help to stimulate creative and appropriately embedded methodological adaptations in readers' own work. Perhaps slightly less obviously, the book's collaborative writing and editing represents an attempt to distanciate (spread and sustain) the politics and practice of PAR beyond place-based field research (Kesby 2007a). Most of the chapters have been produced collectively, with joint or multiple authors (including many non-academics). They also include many text boxes in which participants and researchers voice illustrative stories as a powerful means of sharing experience and effecting change. Through these alternative modes of representation and collaborative modes of writing, we attempt to provoke a questioning of mainstream academic and corporate publishing practice (see also Cahill and Torre, Chapter 23 in this volume), and further emphasise that knowledge production (in texts not just the field) can be a collective participatory activity.

## Overview

The various contributors bring to their chapters rich and diverse insights from rural and urban contexts in Europe, North America, South America, Africa, Australasia and the Pacific. Frequently, ideas raised in one chapter are echoed or complemented by those in other chapters so readers will find many cross-references and common themes to pursue.

**Part I: Reflection** consists of five chapters, which provide the intellectual, ethical and pragmatic contexts for the subsequent chapters exploring 'real-world' applications. Collectively, these chapters foreground several issues, which resurface throughout the book, associated with the spatial politics of practising participation. Chapter 2 clarifies the origins, epistemology and common methods associated with participatory approaches. Chapter 3 examines recent poststructuralist critiques of participatory development and research as a form of power. Chapter 4 considers the difference PAR makes to theory, practice and action. Chapter 5 outlines in detail the ethics of participation (which underpins the rest of the book). Chapter 6 addresses issues of safety and well-being, which are an important, but commonly neglected, part of preparing to embark on PAR.

**Part II: Action** enters into a critical exploration of the politics and practices of PAR. Contributors to the thirteen chapters in this part of the book focus on their engagements firstly with differently situated groups and issues, and secondly with particular approaches and methods. Chapters 7 and 8 discuss participatory attempts to inform environmentally sustainable practices and to secure family livelihoods. They highlight the challenges of scaling up participatory processes and actions. Chapters 9 and 10 address the value of participatory modes of research for topics which can be sensitive and personal. Chapters 11, 12 and 13 investigate issues of approach and design of research for use with specific communities; people with disabilities, migrant groups, and young people. The authors here raise important issues of political representation and reciprocity across different spaces, as well as issues of method.

Chapter 14 provides a critical reassessment of participatory diagramming. A nuanced analysis of participatory cartographies is provided in Chapter 15, which shifts it beyond the realm of 'mapping'. Chapters 16 to 19 discuss participatory art, theatre, photography and video respectively. As emerging and powerful methods which connect affect with effect, they are often able to distanciate the effects of participatory enquiry quite successfully. Finally in this section, Chapter 20 reflects on participatory GIS and the potential for harnessing what have traditionally been seen as top-down spatial technologies in ways that can serve the common interests of communities and researchers.

**Part III: Reflection** considers wider, and often overlooked, issues associated with PAR, in light of the preceding chapters. Chapter 21 attends explicitly to the questions around how we analyse the products and processes of PAR. Chapter 22 provides some frameworks for considering how learning may take place in PAR and how it may be assessed to inform the process. In Chapter 23, the challenging issues of representation and dissemination are discussed in the context of PAR's mandate to inform change by reaching multiple audiences. Some of the institutional benefits and challenges of doing three different forms of PAR within the academy are analysed in Chapter 24. Finally, Chapter 25 returns us to the heart of PAR – the relationship between research, action and change – by questioning the place of activism in our work.

Our final chapter draws together key themes from the book, and reasserts the importance of paying attention to the spatialities of PAR. We highlight a range of scales in which future actions can be focused to further PAR's effectiveness and move us towards more empowering geographies.

# Part I

# Reflection

# 2 Participatory Action Research

## Origins, approaches and methods

*Sara Kindon, Rachel Pain and Mike Kesby*

### Introduction

Participatory Action Research (PAR) has been defined as a collaborative process of research, education and action (Hall 1981) explicitly oriented towards social transformation (McTaggart 1997). It represents a major epistemological challenge to mainstream research traditions in the social and environmental sciences. The latter assume knowledge to reside in the formal institutions of academia and policy, and often presuppose an objective reality that can be measured, analysed and predicted by suitably qualified individuals. In contrast, Participatory Action Researchers recognise the existence of a plurality of knowledges in a variety of institutions and locations. In particular, they assume that 'those who have been *most* systematically excluded, oppressed or denied carry specifically revealing wisdom about the history, structure, consequences and the fracture points in unjust social arrangements' (Fine forthcoming). PAR therefore represents a counter-hegemonic approach to knowledge production.

Various strands of Participatory and Action Research approaches have been practised since the mid-1940s. Worldwide, there exists a strong network of individuals and organisations involved in the theoretical and methodological subtleties of affecting constructive change through research, learning and action. They provide a dynamic and vibrant context for the work reflected in this book. In this chapter, we provide a history of Participatory Action Research's origins and definitions, drawing attention to the heterogeneity of forms, epistemological stances and politics in action in different parts of the world. We also outline the action–reflection cycle typical of participatory and action research processes before briefly discussing frequently used methods.

### Origins

There are several interpretations of PAR's origins and history (Brydon-Miller 2001; Brydon-Miller *et al.* 2003; Brydon-Miller *et al.* 2004; Fals-Borda 2006a, 2006b; Fine forthcoming; Hall 2005; McTaggart 1997; Park *et al.* 1993). We offer our own here. In the post-war USA, Kurt Lewin (1946) coined the term 'Action Research' to describe a research process in which 'theory would be developed and tested by practical interventions and action; that there would be consistency between project means and desired ends; and that ends and means were grounded in guidelines established by the host community' (Stull and Schensul 1987 cited in Fox 2003: 88). He specifically highlighted the 'iterative process of interplay between researcher and participants in which activities shift between action and reflection' (Fisher and Ball 2003: 209–10) – now often referred to as the iterative cycle of action and reflection, or 'spiral science'. Around the same time, Sol Tax and William Foote Whyte began practising research that enabled local people to directly voice their concerns without mediation by an outside expert (Grillo 2002).

In Brazil in the 1960s and 1970s, emancipatory educator Paulo Freire (1972) developed community-based research processes to support people's participation in knowledge production and social transformation. He was particularly interested in the processes of *conscientização* (conscientization) through which poor and marginalised groups developed a heightened awareness of the forces affecting their lives, and then used this greater awareness as a catalyst to inform their political action. His ideas connected with others in the majority world dissatisfied with the ongoing legacies of colonisation, modernistic development interventions and positivistic research paradigms promoted by university-based researchers.

The early 1970s saw the proliferation of Participatory and Participatory Action Research approaches particularly in Africa, India and Latin America. This work represented a new epistemology of practice grounded in people's struggles and local knowledges, but one that reflected earlier movements in India with Mahatma Gandhi. His method of non-cooperation and passive resistance enabled what he called the practice of 'soul power' as people drew on their own knowledges to voice their concerns and actively resist British colonial rule (Sivananda 2007).[1] In Tanzania, Marja-Liisa Swantz has been identified as being the first to use the term 'Participatory Research' to describe her work integrating the knowledge and expertise of community members into locally controlled development projects (Hall 2005). In India, Rajesh Tandon named a similar approach he developed 'Community-based Research' (Hall 1997; see also Brown and Tandon 1983). In Colombia, Orlando Fals-Borda and others were engaged in what they termed 'Participatory Action Research' which sought to develop alternative institutions and procedures for research that could be emancipatory and foster radical social change (see Lykes 2001b).

This first wave of PAR was followed by a second wave in the 1980s, particularly in community development and international development contexts. This wave was most noticeable in the form of Rapid and Participatory Rural Appraisal (RRA and PRA) approaches created as alternatives to cumbersome development surveys and as a means to involve people as agents of their own development (Chambers 1994). By the 1990s, PAR gained greater popularity within minority world institutions and here it blended with strands of Action Research and critical social science (see for example Horton 1993; Park *et al.* 1993; Whyte 1991). Today, Action Research, Participatory Action Research and Action Learning are the most common terms used to describe research that involves:

> a participatory, democratic process concerned with developing practical knowing in the pursuit of worthwhile human purposes, grounded in a participatory worldview… [and bringing] together action and reflection, theory and practice, in participation with others in the pursuit of practical issues of concern to people, and more generally the flourishing of individual persons and communities.
>
> Reason and Bradbury 2006: 1

Some authors continue to distinguish Participatory Research from Action Research, suggesting it is more focused on learning as a vehicle for increasing citizen voice and power in a wide range of contexts, while Action Research is more focused on social action, policy reform or other types of social or systemic change (Taylor *et al.* 2004). For us the distinction revolves around the politics of the research process itself: Action Research does not necessarily engage participants directly in the research process. It can, for example, be an inquiry into one's own life and professional practice with a view to affecting change or institutional reform.[2] Participatory Research and PAR on the other hand, strive to embody 'a democratic commitment to break the monopoly on who holds knowledge and for whom social research should be undertaken' (Fine, forthcoming) by explicitly collaborating with marginalised or 'vulnerable' others (see also Rahman 1985).

Current theory and practice within PAR derive from multiple strands around the world. While there remain considerable differences, methodologically, epistemologically and politically between these strands (Greenwood 2004), there is also overlap between the terms researchers use to communicate their action-oriented research practice, and distinctions are becoming increasingly blurred. However, researchers often add *Participatory* to Action Research to signal a political commitment, collaborative processes and participatory worldview (Reason and Bradbury 2006) that distinguish it from the 'miscellaneous array … [of research] that attempts to inform action in some way' (McTaggart 1997: 1). For many advocates it seems less necessary to add the word *Action* to Participatory Research because emancipation and transformation have always been at its heart (McTaggart 1997). For this reason, Fals-Borda (2006a) uses the acronym P(A)R. However, given rising concerns that participation's institutionalisation within development practice has led to its de-radicalisation (Cooke and Kothari 2001a; Williams 2004a; see also Kesby *et al.*, Chapter 3 in this volume), inserting the term *Action* serves as an important reminder 'that it is participants' own activities which are meant to be informed by the ongoing inquiry' (McTaggart 1997: 2; see also Chatterton *et al.*, Chapter 25 in this volume).

It is also important to acknowledge the distinctive contribution of Feminist Participatory Action Research, because there has been a failure in some accounts of PAR to acknowledge its insights and role (see criticisms by: Maguire 2000; Swantz 1999 cited in Maguire *et al.* 2004; Wadsworth 1999 cited in Maguire *et al.* 2004; Wadsworth 1997). Nevertheless, feminist perspectives shape many chapters in this book. Advocates not only need to be aware of gendered divisions among participants, but also of the potentially gendering effects of poorly conceived PAR practice (first identified by Maguire 1987). Moreover, a feminist appreciation of social inequality as well as the masculinist nature of 'research as usual' speaks directly to the need for collaborative, participatory research. PAR has benefited greatly from feminist insights in terms of epistemological critiques, the development of alternative methods and the traditional commitment to activism within and outside the university (Greenwood 2004).

There are currently several 'schools' within the broad range of academic and community researchers engaging with various forms of PAR, (Fals-Borda 2006a). These have emerged out of particular intellectual traditions and contexts (Box 2.1). While diverse and overlapping, they reflect PAR's emerging geographies and its embeddedness within the communities, environments and institutions in which it takes place.

This diversity of terminology and approaches is both appropriate and challenging. Epistemologically and methodologically, such diversity accords with

**Box 2.1 Some current schools of PAR**

Action Research: Cornell University, USA (Greenwood)
Action Research: Scandinavia (Gustavsen)
Action Research: Austria (Schratz)
Action Learning: Australia (McTaggart)
Participatory Research: International Council for Adult Education and Ontario Institute for Studies in Education, Toronto (Hall)
Participatory Action Research: Germany (Tillman)
Participatory Action Research: Peru (Salas)
Participatory Action Research: Colombia (Fals-Borda)
Participatory Action Research: India (Rahman)
Participatory Action Research: USA (Park, Whyte)
Participatory Action Research: University of Calgary, Canada (Pyrch)
Feminist Participatory Action Research: USA (Brydon-Miller, Maguire)
Participatory Community Research: USA (Taylor, Jason, Zimmerman)
Community-based Research: India (Tandon)
Community-based Participatory Research: USA (Stoeker)
Tribal Participatory Research: American Indian and Alaskan Native Communities, USA (Fisher and Ball)
Constructionist Research: University of Texas, USA (Lincoln)
Participatory Learning and Action (PLA): University of Sussex, UK (Chambers)
Cooperative Research: University of Bath, UK (Reason)
Participatory Learning and Action (PLA): MYRADA, India (Shah)
Critical Systems Theory: University of Hull, UK (Hood)

Source: Brydon-Miller *et al.* 2003; Fals-Borda 2006; Authors' own analysis

PAR's commitment to locally-appropriate public engagement. Politically, it might limit PAR's collective reach and action. However, a number of characteristics of PAR and PAR-researchers are distinguishable (Box 2.2), which are common to most of the various strands. These, as much as any definition, provide a guide to the politics, practice and professional identity of those involved.

## Epistemology and orientation

Any discussion of PAR's characteristics and the researchers involved in it inevitably engages with questions of worldview and epistemology. Reason and Bradbury (2006: 7) argue that to undertake PAR researchers must adopt a participatory perspective or worldview, which 'asks us to be both situated and reflexive, to be explicit about the perspective from which knowledge is created, to see inquiry as a process of coming to know, serving the democratic, practical ethos of action research.'[3]

For us, the key is an ontology that suggests that human beings are dynamic agents capable of reflexivity and self-change, and an epistemology that accommodates the reflexive capacities of human beings within the research process (see also Kesby and Gwanzura-Ottemoller, Chapter 9 in this volume). In Reason and Bradbury's terms, this perspective represents an 'extended epistemology' which draws on diverse forms of knowing to inform action. Such an epistemology represents a challenge to scientific positivism and seeks to practise the radical, suggesting that it is not enough to understand the world, but that one has to change it for the better. PAR therefore emphasises that there is a socially constructed reality within which multiple interpretations of a single phenomenon are possible by both researchers and participants (Greenwood and Levin 1998). Such a perspective opens up spaces for different forms of knowledge generation through methodological innovation and political action.

Not surprisingly, practitioners of PAR currently engage a range of theoretical sources including feminism, poststructuralism, Marxism and critical theory as they take shape through pragmatic psychology, critical thinking, practices of democracy, liberationist thought, humanist and transpersonal psychology, constructionist theory, systems thinking, critical race theory and complexity theory (Brydon-Miller 2001; Cameron and Gibson 2005; Fals-Borda 2006a; Kesby 2005). This diverse base in radical theory, the conceptual contributions and constant self-critique of PAR inherent in its iterative cycles of reflection and action (see Kesby *et al.*, Chapter 3 and Pain *et al.*, Chapter 4 in this volume), illustrate that PAR is not just 'another method'. Rather, PAR is an 'orientation to inquiry' (Reason 2004; see also Kesby *et al.* 2005) which demands methodological innovation if it is to adapt and respond to the needs of specific contexts, research questions or problems, and the relationships between researchers and research participants. PAR also values the processes of research as much as the products, so that its 'success' rests not only on the quality of information generated, but also on the extent to which skills, knowledge and participants' capacities are developed through the research experience (Cornwall and Jewkes 1995; Kesby *et al.* 2005; Maguire 1987). Figure 2.1 provides one example of a typical PAR process.

**Box 2.2 Key characteristics of PAR and PAR researchers**

**Participatory Action Research**

- Aims 'to change practices, social structures, and social media which maintain irrationality, injustice and unsatisfying forms of existence' (McTaggart 1997 cited in Reason and Bradbury 2006: 1)
- Treats participants as competent and reflexive agents capable of participating in all aspects of the research process
- Is context-bound and addresses real-life problems
- Integrates values and beliefs that are indigenous to the community into the central core of interventions and outcome variables
- Involves participants and researchers in collaborative processes for generating knowledge
- Treats diverse experiences within a community as an opportunity to enrich the research process
- Leads to the construction of new meanings through reflections on action
- Measures the credibility/validity of knowledge derived from the process according to whether the resulting action solves problems for the people involved and increases community self-determination

**Participatory Action Researchers are generally:**

- Hybrids of scholar/activist where neither is privileged
- Interdisciplinary
- Mavericks/heretics
- Patient
- Optimistic, believing in the possibility of change
- Sociable and collaborative
- Practical and concerned with achieving real outcomes with real people
- Able to be flexible and accommodate chaos, uncertainty and messiness; able to tolerate paradoxes and puzzles and sense their beauty and humour
- Attracted to complex, multi-dimensional, intractable, dynamic problems that can only be partially addressed and partially resolved
- Engaged in embodied and emotional intellectual practice.

Source: adapted from Brydon-Miller *et al.* 2003; Fisher and Ball 2003; Greenwood and Levin 1998; Greenwood *et al.* 1993; McTaggart 1997; Park *et al.* 1993; Reason and Bradbury 2006

While within PAR, collaboration at all stages of reflection and action is ideal, it is important to recognise that levels of participation by co-researchers and participants may vary significantly. Various authors have attempted to identify different modes and levels of participation, frequently arguing that collective action and self-mobilisation by participants demonstrate successful PAR. Arnstein's (1969) ladder of citizen participation was the first attempt to devise such a schema: the bottom rung being the manipulation of citizens, and the top rung citizen control (representing full and meaningful participation). Since then, others like Hart (1992) have adapted this into the specific context of Participatory Research with children and young people. Here, however, Hart recognised that rather than child mobilisation being the most successful outcome of a participatory process, adults in positions of greater power and authority need to actively support child-led initiatives in a spirit of ongoing collaboration. More recently, Pretty *et al.* (1995) have proposed a participation

| *Phase* | *Activities* |
|---|---|
| Action | Establish relationships and common agenda between all stakeholders<br>Collaboratively scope issues and information<br>Agree on time-frame |
| Reflection | On research design, ethics, power relations, knowledge construction process, representation and accountability |
| Action | Build relationships<br>Identify roles, responsibilities and ethics procedures<br>Establish a Memorandum of Understanding<br>Collaboratively design research process and tools<br>Discuss and identify desired action outcomes |
| Reflection | On research questions, design, working relationships and information requirements |
| Action | Work together to implement research process and undertake data collection<br>Enable participation of others<br>Collaboratively analyse information generated<br>Begin planning action together |
| Reflection | On research process<br>Evaluate participation and representation of others<br>Assess need for further research and/or various action options |
| Action | Plan research-informed action which may include feedback to participants and influential other |
| Reflection | Evaluate action and process as a whole |
| Action | Identify options for further participatory research and action with or without academic researchers |

*Figure 2.1* Key stages in a typical PAR process

Source: Kindon 2005; Parkes and Panelli 2001; and authors' own experiences

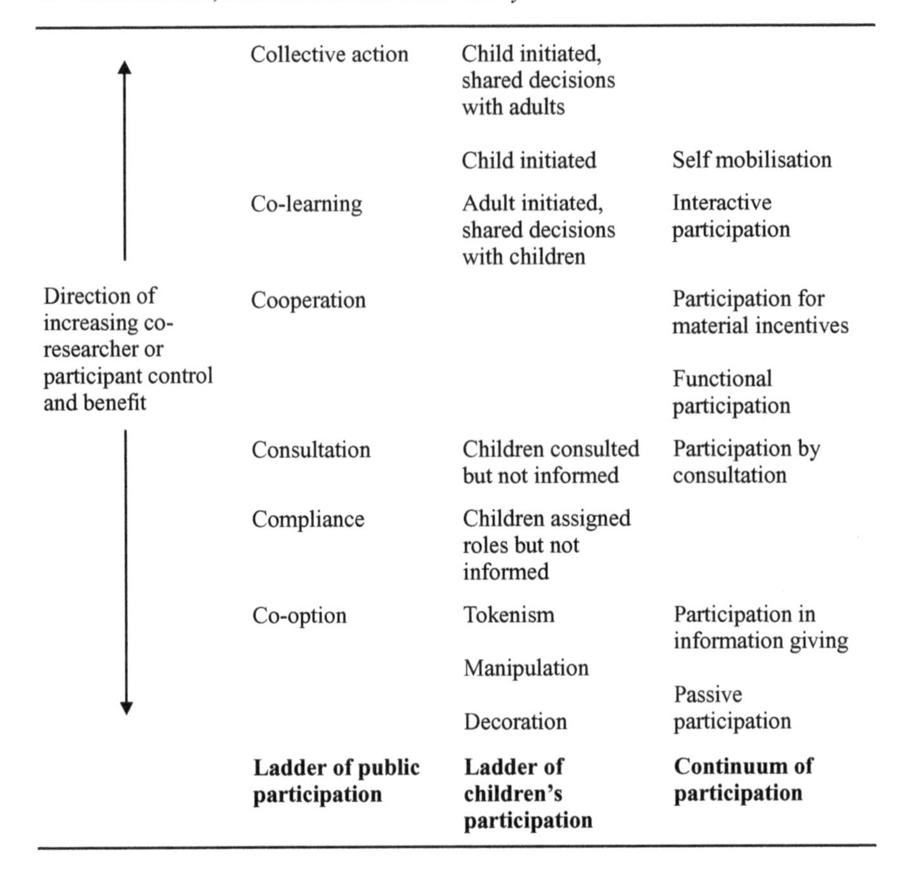


*Figure 2.2* Participation continuums

Source: Arnstein, 1969; Hart 1992; Pretty *et al.* 1995.

continuum, which tries to avoid the fixed value judgments implicit in the ladder's hierarchy, and to acknowledge that various forms of participation may be valid at different times during a research process and in different situations and contexts (Figure 2.2). Clearly, within PAR, choices about modes and degrees of participation are not just made by the researcher but negotiated with co-researchers and participants. The latter may not desire full participation and care needs to be taken to work with people on their own terms (Kitchin 2001).

## Methods and techniques

The most common methods used in PAR focus on dialogue, storytelling and collective action (Figure 2.3). In the last twenty years with falling technology costs, arts and media-based methods have become popular, as have visualisation techniques such as participatory diagramming and mapping where participants

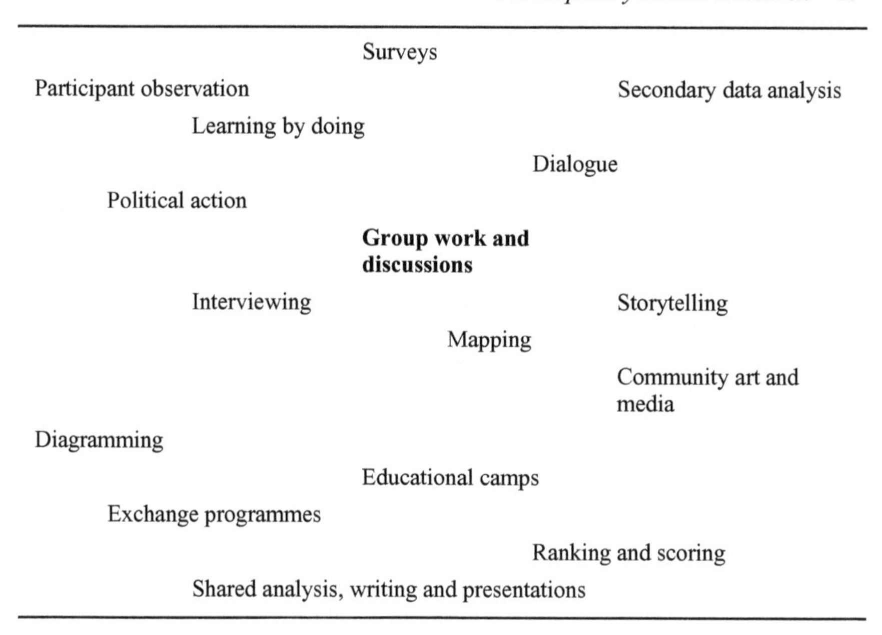


*Figure 2.3* Common methods used in PAR

create charts, pictures and maps to explore issues and relationships (see Chapters 8, 9, 10, 14–20 in this volume). In work with marginalised or vulnerable people, one of the most important features of these types of method is their 'hands-on' nature, and their ability to enable people to generate information and share knowledge on their own terms using their own symbols, language or art forms.

These methods challenge more conventional social science approaches where the outside researcher sets the agenda, decides on the questions to be asked and implements the interview or questionnaire survey for later analysis. The participatory methods and techniques now commonly deployed emphasise shared learning (see Stuttaford and Coe, Chapter 22 in this volume), shared knowledge, and flexible yet structured collaborative analysis (Cahill, Chapter 23 in this volume; Kindon 2005; PLA notes 2003; Pretty *et al.* 1995). They require the researcher to relinquish control (Sense 2006) and take more of a back seat as a facilitator rather than director of the process (Wadsworth 2006). When facilitated appropriately, methods within PAR embody the process of transformative reflexivity in which both researcher and participants reflect on their (mis)understandings and negotiate the meanings of the information generated together (see Crang 2003). This includes paying increasing attention to their changing positionalities and subjectivities throughout the research process (see Browne *et al.* forthcoming; Cahill 2007b; Cameron and Gibson 2005).

Regrettably, there are still many cases of participatory techniques being deployed without a wider collaborative approach or participatory worldview

(Reason and Bradbury 2006; see also Alexander *et al.*, Chapter 14 in this volume). These kinds of deployment in development or academic research can have disturbing political consequences, potentially furthering the scope and technologies of neoliberal power. We explore these effects further in Chapter 3.

## Summary

In this chapter we have emphasised the positive aspects of PAR, in terms of its intended political and material outcomes and its potential to transform unequal power structures and relationships to research and knowledge production. It is these aspects that drew us to PAR. It has not always proven an easy path to follow and, like many of our contributors, we have encountered many barriers to adopting a PAR approach within university environments (see also Fine, forthcoming). Nevertheless, as the many rich and exciting examples presented in this book show, it is a path well worth taking. While we retain a critical and reflexive perspective on PAR and recognise its limits and dangers (Kesby *et al.* 2005; Kesby *et al.*, Chapter 3 in this volume) we remain optimistic about PAR's potential to offer politically engaged and practically possible ways to 'do research differently' that engage and benefit those with whom we work (Pain *et al.*, Chapter 4 in this volume).

## Notes

1 We are grateful to Janet Townsend for reminding us of this connection.
2 Reason and Bradbury (2006: xxvi) distinguish between three forms of AR: First person AR mentioned here; second person AR which involves face-to-face inquiry with others into issues of mutual concern; and third person AR which aims to create a wider community of inquiry involving persons not known face-to-face. They also say that the most compelling AR involves all three.
3 Reason and Bradbury (2006) offer an extended discussion of the 'participatory perspective' they are advancing. We encourage readers to consult their work.

# 3 Participation as a form of power

## Retheorising empowerment and spatialising Participatory Action Research

*Mike Kesby, Sara Kindon and Rachel Pain*

### Introduction

As advocates of Participatory Action Research (PAR), we want to inspire readers to consider how PAR can enrich scholarship and facilitate political engagement beyond the spaces of its immediate intervention. To do this, in the current climate, means grappling with a series of complex and painful critiques, mainly from development studies, that have thrown some doubt on the utility and legitimacy of participatory and action-oriented approaches (notably Cooke and Kothari 2001b). These critiques engage with poststructuralist theories, most notably the work of French philosopher Michel Foucault, and point towards the negative effects of PAR as it has been commonly deployed within the international development context.[1]

In this chapter we acknowledge the value of these critiques and use them to re-engage with the issue at the very heart of PAR – power. Our argument is that while PAR is a form of power, its effects are not only negative. Rather they are messy, entangled, highly variable and contingent. We propose that PAR can better fulfil its potential to effect empowerment if it is reconceptualised as a spatial practice, and if this reconceptualisation infuses practice. We are confident that this shift could enable us to more adequately consider the socio-spatial interplay of the people and settings involved, the resources mobilised and the reach or spread of PAR's effects.[2]

### Participatory Action Research as a form of power

#### *The prevailing poststructuralist critique*

> The core of [poststructuralist philosophy] is the doubt that any method … theory … or … tradition … has a universal … claim to the 'right' or the privileged form of … knowledge …. [Poststructuralism] suspects all truth claims of masking and serving particular interests.
>
> Richardson 1993: 517

At first glance, PAR advocates may be comfortable with Richardson's analysis since, ideally, they strive to diminish their own expert status, valorise the expertise and perspectives of participants, and recognise and subvert the power relations that usually structure research (the imposition of agendas, extraction of data and absence of benefit for researched communities) (Bryden-Miller *et al.* 2003). However, discomfort soon emerges when we are told that, far from circumventing power, participatory approaches (even when done 'properly'/deeply) are themselves a form of power which differs little from other externally imposed forms of research, and should therefore only be resisted (Cooke and Kothari 2001b; Kapoor 2005; Kothari 2001).

Given everything that PAR stands for, how can this state of affairs be so? The issue turns on definitions of power. Poststructuralists do not see power as a commodity that can be held or redistributed, but as an *effect*: an action, behaviour or imagination brought into being in a specific context as the result of the interplay of various communicative and material recourses (see Allen 2003; Foucault 1977; 1978). The resources of participation include various discourses/practices like equity, democracy, collective action, self-reflection and dialogue, as well as the growing consensus around the utility and legitimacy of participatory approaches and techniques. But we need to ask: What are the power *effects* that these resources produce?

To date, the focus of critique has centred on the 'negative' power effects of participation (see Box 3.1). But while there is much insight here, there is also conflation, and we wish to argue that while the resources of participation can and do produce negative effects in some circumstances, these same resources can be deployed to achieve quite different effects (see Foucault 1978). Thus, the effects of institutions like the World Bank deploying PAR are not necessarily the same as those sought or achieved by radical academics, or community activists. Moreover, poststructuralism can itself be a resource that researchers might deploy in the service of participatory approaches (see for example, Cameron and Gibson 2005; Hickey and Mohan 2004; Kesby 2005; 2007a; see also Pain *et al.*, Chapter 4 in this volume).

### *A more nuanced understanding of power and empowerment within Participatory Action Research*

Perhaps surprisingly, we find hope for PAR in the notion of *governance* (see Clayton 2000a; 2000b; Foucault 1991). PAR effects governance because it seeks to proscribe and delineate possibilities for behaviour. Researchers deploy discursive resources such as 'equality', and micro-technologies such as facilitation, to induce participants to reconstitute themselves as reflective agents engaged in a programme of critical self-regulation and analysis. These governing effects of participation are certainly *power-full* – involving or constituting power – but they can be positive as well as negative.

Existing critiques slip too readily between power and domination as if they were the same thing. Yet even when thinking of PAR's 'negative' effects we prefer to

**Box 3.1 Some negative power effects of participatory approaches**

- De-legitimisation of research methods that are *not* participatory
- Production of participants as subjects *requiring* 'research'/'development'
- Production of suitably disciplined subjects as *participants* expected to perform appropriately within participatory processes
- Retention of researchers' control whilst presenting them as benign arbiters of neutral or benevolent processes
- Re-authorisation of researchers as experts *in* participatory approaches
- Romanticisation or marginalisation of local knowledge produced through participatory processes
- Reinforcement of pre-existing power hierarchies among participating communities
- Legitimisation of elite local knowledge simply *because* it is produced through participatory processes
- Legitimisation of neoliberal programmes and institutions (such as the World Bank) that also deploy participatory approaches and/or techniques

Source: Cooke and Kothari 2001; Cornwall and Brock 2005; Henkel and Stirrat 2001; Kapoor 2002; 2005; Kothari 2001; 2005: Mohan 2001; 2007; Mosse 1994; Sanderson and Kindon 2004)

draw on Allen's (2003) more subtle understanding of power's various *modalities*. These interplay, overlap and produce effects in quite different ways. For example, *domination* may be evident when facilitators use 'ground rules' to impose a particular form of conduct on participants, or when participatory techniques impose a particular form of representation on local knowledge (see also Alexander *et al.*, Chapter 14 in this volume). Indirect *coercion* may occur where participation in an intervention offers the only possible hope against the threat of poverty. *Inducement* and *seduction* may be at play where access to resources and skills is promised and aspirations tapped to ensure recruitment. *Manipulation* may occur where researchers use peer researchers to circumvent communities' distrust of academics, or draw elites into self-critique under the cover of seemingly innocuous topics. Finally, *authority* (which facilitates the other modalities mentioned here) may be at work where participants concede an expert status to researchers (whether or not researchers want or realise this).

Clearly there is more to power in PAR than domination. Moreover, the effects of such governance can arguably be positive: in socially divided communities PAR can delineate modes of behaviour that enable unequal agents to participate with equivalence while discussing controversial issues (see for example, Cieri and McCauley, Chapter 17 in this volume). So perhaps resistance is not the only possible response to the realisation that PAR is a mode of power. Allen (2003) offers two further modalities, and reworks another in ways that might describe these positive PAR effects: *negotiation* between unequally positioned participants in pursuit of common goals; *persuasion* by strength of argument in an atmosphere of equality; and *authority among* participants – conceded to researchers who acknowledge their own uncertainty and situatedness.

Conversely, empowerment is not only positive (productive and enabling); it also has negative effects: closing down possibilities (for acting at other people's expense), constraining (the ability to exert power over others) and causing people to behave in particular ways (finding a voice and acting assertively).[3] Indeed rather than releasing an authentic freewill crushed by power, empowerment produces agency in a new 'associational' guise. Empowerment must therefore be reconceptualised as an effect of a form of *governance* that enables people to forge a *common will* and work with others via *negotiation* and *persuasion*.[4]

We argue that imagining empowerment's effects and modalities as overlapping and entangled (Sharp *et al.* 2000) with those of power enables a much more constructive critique of PAR. As PAR researchers, we can identify and attempt to moderate instances of power's more negative effects whilst acknowledging parallel instances of its positive effects: for example, recognising that the imposition of 'ground rules' might simultaneously effect respectful interactions and negotiation between unequal parties, and that manipulation of discussion might enable marginal groups to persuade elites of the need to reconstitute norms. Finally, we could anticipate participants' and co-researchers' resistance to PAR as a form of power and allow their critique to strengthen our practice.

## Spatialising Participatory Action Research

In addition to the points already made, as geographers we believe that PAR can better fulfil its potential to effect empowerment if it is understood as a spatial practice. Interest in the spatiality of participation is growing, but careful thought is needed if a spatial turn is to remain consistent with a poststructuralist critique of participation.

In development studies, the work of Cornwall (2002; 2004a; 2004b) has generated interest in spatial perspectives on participation. She suggests an innovative descriptive taxonomy of 'invited' versus 'popular' the spaces in which participation may take place. However, for us, this framework lacks analytical utility because it is too polarised, harbours conventional views of power and empowerment, and leaves advocates confused about exactly where PAR is located in its imagined landscape of participation.

We agree that there are many instances where supranational agencies, governments and non-governmental organisations (NGOs) 'invite' people into predesigned participatory spaces configured in ways that domesticate their initiative and co-opt them into supporting the status quo. Certainly, these arenas require reform and redesign if PAR is to have any chance of producing empowering effects. However, we disagree that it is the 'invitedness' of a space that makes social interactions within it conservative and/or reactionary. Our experience of

PAR initiatives (as many of the chapters in this volume also show) suggests that 'invited' spaces can facilitate positive interactions and radical transformations.

We also acknowledge that organically emerging 'popular' spaces, fashioned and claimed by people themselves, may offer opportunities for popular education and consciousness raising (see also Chatterton *et al.*, Chapter 25 in this volume). However, we do not agree that it is necessarily the 'popular' constitution of a space that makes social interactions within it radical. Historically, many 'popular' spaces and arenas have been hierarchically structured, authoritarian or motivated by conservative social agendas.

The problem with this formulation is its tendency to see 'popular' spaces as distanced from power. While Cornwall (2002) remains committed to participatory approaches, she discourages 'external interference' (2004b: 82–7) in 'popular' spaces and is unclear whether PAR has a role there. Meanwhile, she fails to explain what alternative resources will constitute these spaces and initiate radicalism therein, or why these would not themselves comprise a form of power. We can only assume the implication is that if participatory approaches are voluntarily adopted within 'popular' spaces (or deployed within reformed 'invited' arenas), that they would not constitute a mode of governance or produce power effects. Thus, Cornwall's new spatial perspective masks an older perspective on power. She attempts to construct an imaginary geography to circumvent the poststructuralist critique and reassert the discredited argument that PAR can, when done properly, be a neutral process that creates and occupies a privileged space, apart from and unsullied by the negative effects of power.

In contrast, our own view is that power and resistance are entangled (Sharp *et al.* 2000) and not just because the one begets the other, or because resistance is often itself associated with authority and domination. The consequences and experiences of governance that effect empowerment may be very different to those that effect power, but the way effects are produced might be seen as similar. Thus, rather than seeking to segregate power and empowerment spatially, it is more productive to conceive the modalities and spaces of power and empowerment as entangled.

## *Re-theorising empowerment and spatialising Participatory Action Research*

The theoretical tiredness of the term empowerment stems not only from attempts to distance and distinguish it from power, but also from the tendency to conceive of it in temporal rather than spatial terms. When defined at all, empowerment is imagined as a more or less linear process of 'enlightenment' (for example see Crawley 1998; Friedmann 1992). However, this formulation does not recognise that agency itself is constituted from available resources or that empowerment is often experienced as hard to maintain over time and/or space. In contrast, by embracing a poststructuralist critique, the concept of empowerment can be theoretically revitalised through an exploration of its spatialities and its similarities with power.

Re-conceptualising empowerment as an *effect* that results from the deployment of certain resources (such as participatory techniques) means that we might *expect* it to be unstable. Judith Butler's (1993) notion of performance suggests that identity and behaviour are brought into being only through repetitive citation of established norms. The epistemologies and methodologies of PAR present useful citational resources upon which participants and researchers can draw to achieve new forms of agency (see Clegg 1989). However, these will need to be constantly redeployed and normalised if empowered performances are to become sustainable. Similarly, if power effects are always already spatial (Allen 2003; Foucault 1984) then so surely are empowerment effects.

Conceiving of PAR as a spatial practice, we see that wherever initiatives occur (in a community centre or under a shady tree) and whether they are 'invited' or 'popular' in origin, they constitute special socio-spatial arenas *governed* by the discourses and practices of participation (such as 'peer equality' and 'facilitation'). Even in meritocratic societies these resources are often significantly different to those normally regulating everyday spaces. In communities sharply structured by social hierarchies they represent radically alternative modes of social interaction and can provide a 'safe space' in which marginal groups can speak and critique everyday society (see also Cahill, Chapter 21 in this volume).

So rather than condemning PAR as a form of power, or seeking to quarantine it from power, we argue for the deployment of poststructuralism in the service of PAR (see also Cameron and Gibson 2005). We suggest that *because* PAR effects governance it can effect empowerment and catalyse radical transformation. The unavoidable paradox is that the governance of participatory spaces can enable the emergence of associational modes of interaction. For example, the authority of those 'inviting' participation, the ground rules that govern behaviour, the manipulations and *seductions* of researchers and so on, can simultaneously facilitate minority voices, enable *negotiation* and *persuasion* rather than domination and submission, and help facilitators to listen to participants. PAR is not the only resource that might produce empowering effects, but more than most, its epistemological orientations and practical techniques can provide mechanisms with which to reflect on its own situatedness and potentially domineering effects.

Nevertheless, the limits that result from PAR's spatial embeddedness need to be recognised. While 'normal' social relations may be partially suspended within PAR arenas, normalcy continues to be reproduced in and through all the (as yet) untransformed spaces surrounding them (see also Cahill and Torre, Chapter 23 in this volume). These spaces and/or relations press in on participatory spaces in a number of ways, potentially curtailing the empowerment effects possible therein. Some groups and individuals will enter participatory spaces better equipped to utilise the opportunities provided (Cornwall 2004b; Jones and SPEECH 2001; Kapoor 2005), and may use them to reauthorise existing social relations (Cooke and Kothari 2001b). Meanwhile, those given new opportunities to speak or act may not seize them for fear that consequences await them elsewhere, in the many everyday spaces over which facilitators' ground rules have no jurisdiction. Finally, we must remember that although PAR opens up new socio-spatial arenas, these emerge within historically produced contexts and landscapes that are already invested with meaning and will be more or less conducive to the deployment of participatory resources (Cornwall 2002; 2004a; 2004b; Jones and SPEECH 2001; Jupp 2007).

A poststructuralist perspective on PAR as a spatial practice offers many useful insights into how and why participation works, and helps to explain why it sometimes fails or is difficult to sustain. As well as seeing empowerment and social change in temporal terms ('Did the period of participation/transformation last long enough?'), we would urge researchers and practitioners to recognise that the empowering effects of PAR are embedded in place ('How far did the environment governed by participation extend?'). But we cannot simply hope that skills developed in one domain of association are transferable to another (e.g. from 'popular' to 'invited' spaces – see Cornwall 2002; 2004a). Consciousness, agency and behaviour are all socio-spatially relational; co-researchers and participants may find it difficult to perform in empowered ways once they leave participatory arenas and return to spaces governed by resources that effect normalcy. If we want our projects to be more than isolated islands of empowerment we need to identify which resources can be successfully redeployed, normalised and distanciated over time–space (see also Cahill and Torre, Chapter 23 in this volume). Only when resources able to effect empowerment are available to ex-participants in other domains of association, will these spaces be reconstituted in ways that make them conducive to the re-performance of empowerment (Kesby 2005; 2007a).

## Conclusion

Acknowledging that PAR is enmeshed with power clarifies how it works as a spatial practice and how empowering effects might be spread and stabilised. We can no longer see PAR as a privileged, power-free mode of research, and must see it as a situated, contestable work in progress. It is nevertheless legitimate (and necessary) to deploy forms of governance like participation to transform more oppressive and less self-reflexive forms of power. PAR can learn theoretically from poststructuralism, and PAR can offer poststructualism a practical means to achieve radical projects of de/reconstruction in and through its praxis.

## Notes

1 We note that these recent critiques seem to overlook earlier critiques by feminists engaging PAR (for example, Maguire 1987).

2 In other articles, Kesby argues these points in more detail (Kesby 2005; 2007a) and Cameron and Gibson (2005) also provide an illustration of how PAR can be carried out in a poststructuralist vein.

3 While we recognise that the goal of securing power *with* rather than *over* makes the political goals of empowerment different to those of power, we feel that attempting always to separate the two is less useful than exploring their similarities (see also Rowlands 1997).

4 Allen (2003) attempts to include empowerment within his poststructuralist discussion, but does not quite manage to do so because he clings to the conventional view that empowerment is entirely different to power.

# 4 Participatory Action Research: making a difference to theory, practice and action

*Rachel Pain, Sara Kindon and Mike Kesby*

## Introduction

This chapter asks how Participatory Action Research (PAR) relates to some current theoretical paradigms, how it informs new research practices, and how it produces change inside and outside of the academy. To explore these questions, we consider PAR's relation to recent social and spatial turns in social and environmental research. As we discussed in Chapter 3, poststructuralist approaches have been most prominent in recent debates about PAR; here we reflect on other significant perspectives. We also identify how our own engagements with PAR in different contexts tackle theory, invigorate practice and open up possibilities for social change and political transformation.

## Engaging theory: PAR and recent social and spatial turns

In Chapter 2 we discussed the historical origins of PAR. More recently, academics have begun to engage with, and contribute to, other bodies of theory informed by feminism, poststructuralism and postcolonialism. These engagements reflect broader epistemological and methodological shifts within the social and environmental sciences, associated with the postmodern questioning of disciplinary truth claims. They also reflect wider societal and political changes associated with increasing globalisation and its related crises of place and identity, the rise of social movements and calls for participatory democracy (Cornwall 2004a; Hickey and Mohan 2004). The so-called cultural turn, materialist turn, and emotional turn witnessed within Human Geography over the past decade, for example, can be seen as reflecting academics' desires to make sense of an apparently more complex world and to make more relevant contributions to it.

One of the longest-standing, if not particularly easy, alliances has been between PAR and feminist theories (see also Kindon *et al.* Chapter 2, and Chatterton *et al.* Chapter 25, in this volume). Feminist principles of equality, reciprocity, partiality and valuing the voices of ordinary people as expert and authoritative on their own lives are reflected in PAR. Since an early critique by Maguire (1987), feminist scholars in Geography have made important contributions to PAR scholarship and practices (see Cahill *et al.* 2004; 2007a; Kindon 1995b; Mountz *et al.* 2003; Pratt *et al.* 1999; Slocum *et al.* 1995; Townsend *et al.* 1995).

More recently, feminist and other scholars have engaged poststructuralist theories, which have stimulated some of the most heated and divisive debates on participation. We discuss their implications for the theorisation and practice of PAR in Chapter 3 (see also Cameron and Gibson 2005; Hickey and Mohan 2004; Kesby 2005; 2007a; Mohan 2007; Williams 2004a).

Postcolonial theories, critical engagements with the legacies of colonialism and the knowledges of indigenous peoples, have also had important dialogues with PAR (Smith 1999; Suchet-Pearson 2004). Postcolonial critiques have suggested that PAR researchers often re-authorise themselves as new experts of participation, rather than truly handing over authority (Kothari 2005); participation thus forms a new kind of subjection (Henkel and Stirrat 2001). However, postcolonial practice of PAR, especially by feminists, has also reworked positive and empowering alternatives (see Kindon 2003; see also Tolia-Kelly Chapter 16, Krieg and Roberts Chapter 18, and Hume-Cook *et al.* Chapter 19, in this volume).

There are some less obvious relationships between PAR and recent shifts within social and spatial theory, often referred to respectively as cultural, material and emotional turns, which we now go on to discuss. Meanwhile, Box 4.1 explores the relationship of PAR to critical Human Geography.

The *cultural turn* of the early 1990s involved a shift towards culture in social theory, focusing on the roles of text, discourse and representation (see Duncan and Ley 1993). Some viewed it as a negative move away from political economy and structuralist approaches, which often raised questions of radical social change. Yet the cultural turn was multifaceted. It developed from and with various feminist,

**Box 4.1 The place of PAR within human geography**

The profile of PAR and its variants is well-established within critical Human Geographies. This booming sub-area incorporates feminist, postcolonial and other radical perspectives in an effort to understand space and society through 'analyses that are critical and radical [and] are understood to be part of the praxis of social and political change aimed at challenging, dismantling, and transforming prevalent relations, systems, and structures of capitalist exploitation, oppression, imperialism, neo-liberalism, national aggression, and environmental destruction' (ACME 2007: Index). Some have expressed a desire for more engagement and commitment to change among critical theorists (Fuller and Kitchin 2004; Kesby 2005; Pain 2003). This has marked a turn (back, and forwards) towards the political potential inherent within participatory approaches. As yet, however, PAR's contribution to the more recent emotional and material turns is only just beginning to be explored.

Source: Authors' own analysis

poststructural and postcolonial theories, and opened up spaces for the voices and histories of marginalised social groups. PAR is a material and political means of progressing this agenda, offering as it does practical deconstruction and reconstruction of academic spaces. Further, the cultural turn was accompanied by a growth in qualitative methodologies (Crang 2003) within which PAR gained greater legitimacy.

More recently, a focus on the roles and effects of emotions has blossomed (see Ahmed 2004; Davidson *et al.* 2005). Many social and spatial processes and institutions, formerly viewed by the social sciences as rational and disembodied, are being re-evaluated in this *emotional turn*. The development of non-representational theory and related theories of affect pay close attention to non-verbal and pre-cognitive knowledge, as well as to bodies as unconscious receivers and transmitters of affect, facilitated by spatial settings (Thrift, forthcoming). These literatures have identified a need for methods that attend to the range of sensory experiences that inform knowledge and action, by reaching beyond the limits of text, the verbal and the material, to engage new media, lived experiences, performative, haptic and embodied knowledges (see Crang 2003; Thrift 1997).

Finally, and partly in response to the perceived over-emphasis on textual analysis apparent within the cultural turn, the recent *material turn* (Whatmore 2006) has re-focused attention upon the materialities of objects within particular cultures and landscapes (see Anderson and Tolia-Kelly 2004; Latham and McCormack 2004). In this shift, epistemological value is given to matter as a means of understanding the relations between people and places, requiring methodological engagements with the non-human world of things. One suggestion is that materiality might be re-mobilised to challenge injustice and ground political struggles (Gregson 2003; Pain 2006).

To date, PAR within social and environmental sciences seems firmly situated within critical arenas (Box 4.1), perhaps reflecting the legacies of its origins within liberal humanist and scientific paradigms. However, PAR has much potential to speak to the shifts in social and spatial theory discussed above, given its emphasis on inclusive methods, non-verbal understandings and visual products. For example, PAR's methodological innovations demonstrate the powerful political effects that visual representations (such as maps, diagrams, photographs, video and art) can have. Moreover PAR, as the contributions to this volume demonstrate, has a long history of drawing on and engaging with emotional and affective registers, representing these in non-verbal and non-textual ways to influence change (see also Cahill 2007b; Cameron and Gibson 2005).

## Making a difference to theory

From our experience, what makes it possible for PAR to make a difference in all these areas of theory is that it works on the basis that there is no singular or fixed version of reality awaiting detection; its very premise is emergence and being open to the many realities 'out there'. Emphasis is placed on collaborative knowledge production and knowledges performed intersubjectively in and through research processes (Gibson-Graham 1994; Jupp 2007; Kesby 2007a). The politics of most PAR practice are never fixed, but are both a politics of *becoming* (Cameron and Gibson 2005; Mountz *et al.* 2003) and *betweenness* where knowledge, analysis and action emerge between co-researchers and participants. The action and push for social change which may emerge from PAR (see Chatterton *et al.*, Chapter 25 in this volume) constitute open and fluid possibilities which rest upon our co-researchers' needs and wishes, reflections and actions.

The centrality of the iterative cycles of reflection and action within PAR (Kindon *et al.*, Chapter 2 in this volume) informs its relationship with theory in three main ways. First, PAR reflects and affects the spaces within which it takes place; in other words, PAR and the spaces with which it intersects are mutually constitutive. Geographers and others have begun to explore these *spatialities* and to acknowledge the important connections between people, participation and place. Their explorations are shedding new light on the relationships between PAR, empowerment and space, as well as theories associated with citizenship, sustainability and social change (see for example, Cornwall 2002; 2004a; Gaventa 2004; Kesby 2007a; Kesby *et al.*, Chapter 3 in this volume; Kindon and Pain 2006; Mohan 2007; Pain and Kindon 2007; Williams 2004a).

Secondly, PAR alters the way that we theorise and produce knowledge within and between these spaces through its attention to dialogue, relationships and inclusive methods. The explicit attention to *relationalities*, as Ellis (2007) outlines, prioritises and values relationships and aims at 'mutual respect, dignity and connectedness between researcher and researched ... [It] requires researchers to act from our hearts and minds, to acknowledge our interpersonal bonds to others' (2007: 4).

Thirdly, and informed by these spatialities and relationalities, PAR leads to new or modified theories of its *materialities* and objects of study, such as environmental degradation, the circulation of social capital, or the impacts of crime in rural areas (Box 4.2). Because knowledge is co-produced with co-researchers and participants in particular places, those with whom we work can profoundly change our research practices and theoretical perspectives. For example, in their work with Australian Aboriginal women, Howitt and Suchet-Pearson (2006) discuss how differences in co-researchers' cultural understandings of environments fundamentally shaped how the research was conceptualised. PAR's grounded and relational orientation towards knowledge construction therefore challenges the hegemonic norms of insular academic production and enables more diverse and contextually-rich theorisations and actions to emerge.

### *Making a difference to practice and action*

As we suggested in Chapter 2, PAR can be defined primarily in terms of its ethical and dialogic engagement and its commitment to research-informed action. With this book we are keen to develop a re-radicalised deployment of PAR, and to distinguish it from both the mere use of alternative techniques and the deployment of the discourse of participation to serve conservative goals (Kesby *et al.* 2005).

**Box 4.2 Making a difference to theory through PAR**

The PAR project I am involved in is exploring the fears and hopes of young people from diverse backgrounds in the cities of north east England. We have examined how emotions are positioned in a geopolitical climate and risk at the local level. Participants are choosing a range of methods to highlight and analyse particular issues, document emotions, disseminate findings and act on these. The project is engaging critically with theories of critical geopolitics, and geographies of emotion and affect. Existing theories of emotion in relation to geopolitical events rarely seem to be informed by the feelings of real people in real places. The issues arising from the research are so pressing that a research approach which documents and theorises, but doesn't find ways to act on the knowledge produced, would be inappropriate.

With respect to how this informs practice, first of all, PAR disrupts researchers' monopoly on possessing and controlling what is ethical, and demands negotiation with co-researchers and participants (Kindon and Latham 2002; Manzo and Brightbill, Chapter 5 in this volume; Sanderson and Kindon 2004). As such, engaging in PAR from within the academy can be hard because inflexible ethics review board structures cannot easily accommodate PAR's fluid and evolving nature (see also Cameron, Chapter 24 in this volume). As several chapters in this book make clear, there are many reasons for challenging institutional and personal practices found within universities. These relate to their growing role as corporate knowledge producers in neoliberal societies (Castree and Sparke 2000).

For many who practice it as an approach to research, the principles of PAR also reach into and connect a wider set of activities – teaching and supervision (see Kindon and Elwood, forthcoming b; Moss, forthcoming), widening access and recruiting students (Hopkins 2006), raising questions about university suppliers and sponsors (Chatterton 2005), and academics' roles more generally as members of their families and communities. Given the emphasis on collaborative research and learning, PAR projects often leave behind enhanced capacities. Because participants have been directly involved in designing and conducting research, they gain and exchange skills, and may continue researching and campaigning with or without the support of an academic (see Pratt *et al.*, Chapter 12 and Elwood *et al.*, Chapter 20 in this volume).

Second, this negotiation and dialogue requires that researchers and 'researched' engage in the practice of inter-reflexivity as a means of attending to the shifting power relations at play and their impacts on the research process and its products (Box 4.3). PAR requires researchers to negotiate changing and fluid understandings of being inside or outside throughout a project's life. It may also require a movement beyond the bounds of 'normal' research relationships into the less clear-cut realms of friendship and kinship (see also McFarlane and Hansen, Chapter 11

**Box 4.3 Making a difference to practice through PAR**

I have found that attempting a PAR approach to my research on sexual health in Zimbabwe has made a considerable difference to my research practice. Initial differences emerged in the area of data collection. I experimented with participative diagramming techniques, which enabled participants to be more engaged with research, literally seeing the results as they materialised in the field and engaging in provisional analysis. This interreflexivity not only makes my published accounts more robust but also helps me to develop a research practice for action not just knowledge production. Going forward I would like to further develop my research practice by engaging participants more fully in research design and by exploring the possibilities for participatory feedback and dissemination.

Source: Mike Kesby

in this volume). The relational aspects of PAR also demand a rethinking of positionality (Browne *et al.* forthcoming). Whatever form these relationships take, personal transformation is almost inevitable and often highly desirable (Cahill 2007a; see also Cameron, Chapter 24 in this volume). That said, Kapoor (2005) reminds us that we need to be vigilant about our own power, and never unquestioningly assume PAR's benevolence.

Third, PAR's commitment to research-informed action demands a different relationship to 'research outputs' and the audiences of our work. While some have criticised the goal of generating products to affect change as instrumentalist (Demeritt 2005), it need not predetermine 'answers' or even the form of change itself, if an exploratory and emergent approach is taken. Academics traditionally add to knowledge through articles in learned journals and books, but any effects on social change are likely to be long term and indirect. As the work of many authors in this volume demonstrates (see especially Cahill and Torre, Chapter 23), PAR emphasises the production of different outputs before journal articles are even conceived (such as community reports, newsletters, presentations, websites, video, drama productions, art exhibitions, training packages and campaign materials).

PAR demands that academics bridge the worlds within and beyond the academy, and they often engage with the policies and publics who are important to resolving the problems and issues which PAR raises. Unfortunately, as Chatterton *et al.* (Chapter 25 in this volume) note, the proponents of so-called 'public sociologies/geographies' have engaged PAR relatively little. Indeed these movements have been overwhelmingly masculine ones, where 'scientists' involved with communities take on high profile roles in communicating findings to a wider public. Attention is now needed to the far greater number of academics who engage in quieter, collaborative processes of change (Pratt 1998).

This point about engaging in processes of action and change is crucial, particularly where PAR researchers work with young people or other (so-labelled)

**Box 4.4 Making a difference to action through PAR**

I am about to meet with service providers from across my university who are committed to implementing the recommendations of a PAR report disseminated last year by three of my graduate students. Their study - the first of its kind in *Aotearoa*/New Zealand - involved a refugee-background co-researcher and participants, and university service providers in a series of participatory workshops and discussions. It seems likely, as result of this process, that next year upon enrolment, students will be able to identify their refugee-background status and access services and programmes specifically designed to support their belonging in the university and their overall academic achievement.

Source: Sara Kindon

'vulnerable' groups (Box 4.4). It might involve becoming activists alongside social change movements and getting involved in advocacy. Forging alliances requires some parity in political goals, and sometimes complex negotiations, but adds to an existing canon of action as well as facilitating the research (Pratt 1999; and see also Pratt *et al.* Chapter 12, and Chatterton *et al.*, Chapter 25, in this volume). Some PAR researchers seek to lobby and influence policy-makers if the findings and participants demand it (see Pain and Francis 2003). Another approach involves working directly with institutions, policy-makers, or public or charitable organisations from the outset (e.g. Fuller *et al.* 2003; see also Cameron, Chapter 24 in this volume). Depending on the sympathy of the organisation towards participatory principles, and their resources, this model can result in more or less 'deep' participation.

## Summary

While often perceived to be separate from recent cultural, emotional and material turns in social theory, we have argued that PAR is well placed to make significant contributions to them and their social relevance. Similarly, feminist, poststructural and postcolonial perspectives offer valuable means through which to tease out PAR's spatialities and to inform the connections between people, participation and place. In the chapters that follow, authors illuminate some of these connections as they discuss their efforts to make a difference to theory, practice and action.

# 5 Toward a participatory ethics

*Lynne C. Manzo and Nathan Brightbill*

## Introduction

Engaging in Participatory Action Research (PAR) and practice raises critical issues for research ethics. These issues emerge in two ways. First, participation pre-supposes a commitment to a set of values that more traditional research and practice do not necessarily require or embrace. These values have distinct implications for ethical decision making and can potentially augment and extend 'classic' ethical principles such as beneficence and respect for persons in ways that enhance research ethics as a whole. Second, the unique and complex characteristics of participatory work raise ethical questions and practical challenges not typically encountered when working within more traditional research paradigms (see also Kindon *et al.*, Chapter 2 in this volume). Advocates of PAR must therefore call for reform of existing institutional ethical review procedures so that they can accommodate rather than frustrate ethical practice in and *through* participatory research.

Ethics are a system of principles and rules that help us determine which actions are right and which are wrong (Beauchamp and Childress 1998). In this chapter we are concerned with applied normative ethics rather than abstract theorising, and consider the practical application of ethical principles in real-world research situations and decision making. In general, making ethical decisions requires determining 'right action' in light of individual and commonly held values, and the formal ethical principles espoused by academic disciplines and the institutions in which we work. In PAR we must interrogate our choices and actions not only against these standard ethical principles, but also against the principles of participation. However, while many advocates of PAR adopt the approach precisely for ethical reasons, doing so does not entirely circumvent ethical dilemmas.

Indeed, PAR can be more riddled with dilemmas than other forms of research: for example, participant anonymity cannot be guaranteed in community group work focused on local change; giving participants a voice risks revealing their survival strategies to those who oppress them; projects can engage ordinary people in potentially controversial social action; and shared control over the research creates ethical conundrums that emerge throughout the process and are not easily predicted at the outset. Current institutional ethical guidelines do not adequately address these unique dynamics. Consequently we need a better understanding of the ethical principles behind participatory work to cope with these particular dilemmas and to develop standards for 'best practice' without losing the unique flexibility of PAR approaches. This will require PAR advocates to present a series of challenges to the current guardians of research ethics.

We begin by reviewing how traditional ethical mandates found in institutional ethical review requirements work in relation to the principles of participation. From there we will identify some of the ethical dimensions of participatory research in an effort to augment and extend current interpretations of research ethics.

## Existing principles governing research ethics

Increasingly academic research must adhere to institutionalised ethical regulations. In the United States these are known as Institutional Review Board (IRB) requirements or 'Human Subjects' review. Ethical regulations within the US academy are predicated upon The Belmont Report, written in 1979 by the National Commission for the Protection of Human Subjects of Biomedical and Behavioural Research.[1] These reports established three primary ethical principles.

- **Respect for persons** People should be treated as autonomous agents, i.e. capable of deliberating and making decisions. Those with 'diminished autonomy', for example, children, prisoners, those with mental disorders, are entitled to protection.
- **Beneficence** This principle focuses not only on doing no harm but maximising beneficial outcomes for participants, society and/or humanity. It requires balancing the potential risks and benefits of research.
- **Justice** In modern democracy, this principle focuses on treating human beings as equals and distributing good equally among them. Thus, research should not be exploitative and there should be a fair distribution of risks and benefits.

These three core ethical principles are certainly applicable to PAR; the problem lies in the way that adherence to these principles is currently interpreted and enforced through a rather inflexible set of institutionalised rules governing academic research. It is here that we can see a divergence between existing ethical review procedures and the ethics of a participatory approach.

The first problem with existing institutional ethical review procedures appears to be practical but is actually rooted in philosophical differences. Advocates of participation get frustrated that IRB-type structures require researchers to design research in its entirety (including all methodologies and tools) before a single participant is recruited. This assumes that research can be fully pre-planned and will progress in a relatively predictable and linear fashion. However, because PAR advocates seek to share control with participants, they know projects can shift unpredictably in response to the social dynamics and changing needs of participants. These aspects make it difficult, even impossible, to declare every ethical issue to a review board before research begins and can frustrate PAR projects and/or reduce the scope for participation within them. Similarly, current procedures do little to facilitate ethical decision making *within* participatory projects. While ethics committees recognise power inequalities between researchers and the researched, they rarely anticipate that the research process might itself transform that relationship. Rather they assume that power differentials will be maintained and the committee's role is simply to protect participants from potentially exploitative researchers. Whatever the long term impacts of knowledge generated in this way, the research process itself does little to advance social justice or encourage ordinary people to actively participate in ethical decision making and praxis.

The second problem with existing review bodies results from the limited way in which most interpret the three core principles that currently guide ethical research. For example, the principle of beneficence is composed of two dimensions – doing good and doing no harm – but most social science review boards interpret beneficence exclusively as an imperative to 'do no harm'. The idea of 'maximising beneficial outcomes' is often by-passed or seen as more relevant for medical research. Thus most social researchers believe themselves to have behaved ethically if they have had no negative or perceivable impact on the people with whom they have worked. By comparison, advocates of PAR seek precisely to 'have an impact', to instigate change and to create benefits for participants. From the outset they attempt to do something more than collect data and leave without a trace. Indeed a PAR-inspired understanding of social justice suggests that it is in fact unethical to look in on circumstances of pain and poverty and yet do nothing. Simply doing no harm not only neglects the second dimension of beneficence but neglects the privileged positionalities that enable researchers to observe but choose not to become 'involved'. Not only does PAR focus on both sides of beneficence, it actively seeks help from participants to generate benefits for themselves.

We are arguing, therefore, that a PAR approach can actually extend the core principles that govern ethical research. Another way in which PAR can do this relates to 'respect for persons'. Currently, this is expressed primarily through rules regulating voluntary informed consent and privacy that stipulate that:

- research participants have the capacity to understand the risks and benefits of participating;
- they have the legal right to be informed of what these are; and
- they have the right to participate only on a voluntarily basis.

In addition, respect for persons is closely linked to respect for privacy, which is protected through procedures that guarantee confidentiality and anonymity (Robson 2002). The current 'best practice' is that research commences with the distribution of consent forms that state the purpose of the research and its potential risks and benefits, which participants must sign to signal their understanding and approval. Forms must specify the steps to be taken to ensure confidentiality and secure storage of the data collected. While advocates of PAR understand and respect the principle of anonymity, it is often in tension with their use of group-based methods in community settings, and with their desire to facilitate marginal peoples' voices in their own community and in broader political processes. Where research is a 'two-way conversation' an insistence on anonymity can muffle the voices of participants while authorising that of the researcher. While these issues can be dealt with in PAR praxis via participatory discussions around ethics, they make it hard for 'full-time researchers'[2] to fill in conventional ethics forms that tend to expect certain responses to questions about whether researchers intend to guarantee the anonymity of 'respondents'. While the traditional model of research assumes a static and hierarchical relationship between the researcher and participants, advocates of PAR want to reinterpret and extend the notion of respect for persons, not only informing participants about all aspects of the research, but doing so in order that they might *themselves* make decisions about how the research should be conducted.

Finally, while the three core principles of contemporary research ethics have made a vital contribution to good research, they have tended to do so at the expense of other, equally valuable principles. Notably, formal ethical review bodies have not incorporated an ethic of care in their governance of researchers' ethical behaviour. An ethic of care is usually associated with feminist scholarship (see Held 1995; Gilligan 1982), and emphasises empathy and relationships rather than objective decisions. Here, ethical decision making and behaviour is rooted in commitment to others. In an ethic of care, morality is seen as enabling effective engagement rather than as a constraint that limits our individual pursuits. This is commensurate with the PAR approach that involves a lessening of distance between the researcher and community members. Few moral theorists stress good relationships in their discussions on morality (Held 1995), but the ethic of care, like participation, considers ethics as relational. Inclusion of such a principle at the foundation of the ethical review process can raise the ethical standard for academic research and PAR is well positioned to promote this ethic.

In summary, we suggest that while IRB regulations impose moral minimums they do not provide sufficient guidance for navigating the ethical and practical complexities associated with a PAR approach. Certainly, rules are necessary to maintain the integrity of both the research and the institutions through which it is conducted, and advocates of PAR cannot assume that they have adequately addressed ethics simply by adopting a participatory approach: for example, participants still need to be given full information about the likely consequences of participating before they consent to do so. However, our contention is not that PAR practitioners should be exempted from formal ethical review, but rather that the principles governing such review should be re-examined and extended in light of PAR theory and practice.

## Toward a participatory ethics

To move toward a participatory ethics, we now attempt to identify some dimensions of participation that have direct implications for ethical decision making. While not necessarily displacing the established importance of 'justice', 'beneficence' and 'respect for persons', these dimensions warrant serious consideration and can provide additional ethical principles against which researchers can test their choices about right action. Underpinning this alternative approach is the belief that research should facilitate the development of an informed critical perspective among participants (not just researchers), a process Freire (1988) famously described as 'conscientisation' (see also Kindon *et al.*, Chapter 2, and Stuttaford and Coe, Chapter 22 in this volume). Ethical behaviour is an important part of such a critical perspective. If in a participatory approach ethics are by nature contextual, relational and dynamic, then choices that might be deemed ethical may vary over time and from project to project as they emerge from debate among all participants.

### *Representation*

Participatory research accepts and embraces the idea that all knowledge is situated, that neutrality is neither possible nor desirable, since we all have some interest or worldview. Research collaboration is rooted in such an awareness, and participants are valued for the array of interests and perspectives they bring to research. Like an increasing number of other social researchers, advocates of PAR question the assumption inherent in research 'as usual' that researchers have the exclusive right to represent respondents. However, they go beyond limited attempts to give the 'other' a 'voice' in academic texts, by recognising participants' ability to represent themselves *throughout* the research process and to help direct that process (see Chapters 7–24 in this volume). Advocates of PAR make a virtue of sharing and clarifying roles, responsibilities and decision making on an ongoing and reiterative basis (see for example Pratt *et al.*, Chapter 12, and Hume-Cook *et al.*, Chapter 19 in this volume). Facilitation of participants' self-representation could be seen as developing existing notions of 'justice' and making an important contribution to research ethics.

### *Accountability*

In the US and elsewhere, all academic researchers are accountable to ethical review board structures, and more generally to their disciplines and their peers, but a PAR approach extends the notion of accountability to a much broader social field. Because participatory researchers engage in collaborative projects as a means to building valid knowledge and fostering positive social change, they have multiple responsibilities and are also accountable (arguably primarily) to their participants, partners and to the communities with which they work. In addition to formal academic ethical review, these stakeholders will decide whether the research is ethically sound, valid and worth doing. Thus the PAR model broadens current interpretations of accountability. Arguably the PAR ethic of demonstrating accountability to researched communities is an example of 'best practice' that institutional ethical review boards should encourage more broadly.

### *Social responsiveness*

Because it is collaborative and change-oriented, PAR requires researchers to be responsive to the needs and perspectives of participants. Researchers and participants recognise that they are perpetually in relation with one another (Noddings 1995). The ethic of social responsiveness produces a research process that is fluid and which changes in response to different situations and the needs of participants. Although the ethic of social responsiveness is connected to the principle of beneficence and respect for persons, it is also not yet encouraged by IRB regulations. In fact, the very shifts in the research process that social responsiveness requires makes PAR incompatible with institutional ethical review requirements that demand we map out research in its entirety before we begin. Thus, PAR workers are left having to weigh the desire to be socially responsive against the requirement to submit complex and time consuming amendments to review boards. For participatory researchers, social responsiveness must take priority because collaboration is sought as a means toward developing solidarity and initiating change. We propose that ethical review procedures be reformed in order that PAR project proposals need not be submitted 'complete' (invariably requiring amendment later), but are allowed to adopt a phased approach. This could begin with an initial application that covers only early community consultation and participatory research design. Alternatively, clearance could be sought for a general set of ethical procedures describing internal *participatory ethical review* within the project. These could be open to Institutional Review Board inspection at any point, and/or to summary audit and review that could establish best practice for future PAR projects.

### *Agency*

PAR further contributes to the development of research ethics because it not only requires ethical behaviour from the researcher, but also extends this requirement to participants themselves. Principally PAR broadens the ethical principle of respect for persons: every participant in a PAR process must accept the responsibility of recognising that each of their peers has a right to a voice and a valuable contribution to make. PAR also extends and circulates the ethical notion of competent agency, particularly via encouraging participants to recognise that they each have the ability to initiate and enact change. Once again, we suggest that in this area review boards might look to PAR for 'best practices' to extend ethical behaviour through research.

### *Reflexivity*

Ethical dilemmas are usually seen diachronically: as a snapshot of a moment, and as something to be predicted and anticipated in advance. However, ethical dilemmas encountered in participation are best understood in a process-oriented way. PAR is reliant upon and seeks to facilitate the competence and reflexivity of participating people. This means that the PAR process is flexible, socially responsive and emergent and so the questions and issues that require ethical decision making only materialise as the collaboration between participants and researcher progresses. Hence, participants should be allowed and encouraged to engage in the ethical review of their own projects. Throughout the many phases of a PAR project, participants must constantly reflect on their beliefs, motivations and actions and ask themselves: 'What kind of change agent am I and how am I accountable for my own actions?' Here again we point to the need for reform in institutional ethical review and the need to allow phased review for PRA projects.

## Conclusion

Participation will not, in and of itself, make research 'ethical'; the approach can be deployed to support a researcher's pre-existing agenda, or to further the interests of a particular group. Arnstein's (1969) 'ladder' warned us long ago that participation may not achieve shared power but can languish at the level of manipulation (see also Kindon *et al.*, Chapter 2 in this volume). Indeed, we would argue that those PAR researchers seeking to achieve higher-order, more collaborative forms of participation should be more aware than most of the importance of the fundamental ethical principles of respect for persons, beneficence and justice enshrined in existing ethical review procedures. However, most contemporary institutional structures offer too little guidance and insufficient room to manoeuvre in PAR projects. This is unfortunate because projects attempting genuinely collaborative forms of participation present complex ethical issues, not least because of the greater number of people involved in the research design and decision making. Genuine participation requires that ethics are a constant concern not easily dispensed with at 'the research design phase' *before* a researcher 'enters the field'. Our concern is that the application of current research models not only stifles the development of a participatory ethics *amongst* the researched, and neglects the possibilities of 'doing some good' *through* research, but it can also lead researchers to approach ethics as a bureaucratic rule-following exercise (see Beauchamp and Childress 1998). Without doubt, rules help ensure that the important but restricted goal of protecting respondents from abuse and exploitation is met, but it misses a broader opportunity to facilitate a wider social discussion and ethical behaviour through research praxis. Ethics as conventionally interpreted tends to buttress existing power relations in research and society and neglect the possibility that research, and research praxis, can contribute to challenging undesirable social phenomena (see Marcuse 1976).

Re-examining established ethical principles in light of the ethics of a participatory approach can extend our understanding of research ethics in general and provide new insights for principled action. In particular this can broaden our appreciation of who can be involved in ethical decision making and to what end ethical research might be applied. Moving ethical principles and ethical review criteria forward will require a collective effort among PAR practitioners. Redefining the ethical principles for research while accounting for flexibility will be necessary in proposing changes to ethics review board structures that can accommodate and value participatory work. We propose that advocates of PAR can begin to act at the scale of individual institutions by proposing pilot projects to explore new ways to conduct and construct ethical review processes. Scaling-up these initiatives can be facilitated by a resolution among colleagues to share the results of these experiments.

May (1996) argued that moral integrity has at least three aspects: mature development of a critical point of view, a coherence of value orientation, and disposition to act in a principled way. Our point is that an ethic of participation and a participatory ethics can enable people other than full-time researchers and institutional ethical reviewers to develop such critical awareness, values and behaviour. However, for this to happen advocates of PAR need to lobby for the reform of existing institutional ethical review systems so that they are more conducive to the conduct of PAR projects. This is necessary if we are to conduct truly emancipatory and ethical research.

## Notes

1 Both the commission and report were created in response to cases where crimes against humanity were perpetrated through research (e.g. experiments on Jews in Nazi concentration camps which resulted in the Nuremberg Code for biomedical research, the Tuskegee syphilis study where medical treatment was purposefully withheld from African-American men who had the disease) and abuses of people through psychological research like Milgram's experiments on obedience, which was initiated, ironically, to understand the blind obedience displayed by Nazi soldiers.

2 We use the term 'full-time' researchers here to distinguish between researchers in academic or research institutions versus community researchers who are often only able to participate part-time.

# 6 Participatory Action Research and researcher safety

*Mags Adams and Gemma Moore*

## Putting safety centre-stage in Participatory Action Research

Instances of physical threat or emotional trauma often go undocumented in academic literature,[1] yet numerous social researchers have found themselves in compromising situations that have threatened their physical and/or emotional well-being. In this chapter we challenge this deficiency by raising awareness of the possible risks associated with research, particularly Participatory Action Research (PAR). We also propose some strategies for researchers to manage these risks as ethically and inclusively as possible. We believe this discussion will be valuable to other researchers (see Punch 1994).

By focussing our attention on research that is primarily considered to be 'safe', we hope to sharpen the awareness of all researchers, whether they consider themselves to be working in 'safe' or potentially 'hostile' environments. We draw on our experience of various projects, including those we have shared since 2004. While neither of us is explicitly a Participatory Action Researcher[2] we have been asked by the editors to bring our knowledge to the PAR field where, though little discussed, safety issues are no less serious. PAR researchers are very attuned to questions of ethics, to the need to change their behaviour in order to facilitate and respect others' knowledge, and like other researchers, seek at the very least to 'do participants no harm' (see Manzo and Brightbill, Chapter 5 in this volume). Moreover, 'conflict management' is a growing concern among PAR researchers conscious of the particular 'hazards' of bringing differently positioned people together in sensitive group discussion. However, these considerations are rarely cast as *safety* issues and the physical and emotional well-being of the researchers themselves is simply assumed rather than addressed. We want to suggest that PAR researchers can, with forethought and training, make assessments of potential conflicts and dangers, thereby minimising the possibility of getting into difficulties, while optimising the outcomes if difficulties do arise.

## Non-threatening or 'safe' research

We focus on situations that involve studies of 'benign' or 'safe' subject matter, not 'dangerous' fieldwork in 'risky environments' where special prudence is necessary (see, for example, Davison 2004; Dowler 2001; Ensign 2003; Jamieson 2000; Lee-Treweek and Linkogle 2000). Highlighting the potential dangers in 'innocuous' settings helps promote the idea that thorough risk assessment is essential in all research.

PAR research is generally envisaged as a safe activity and any safety issues are usually directed at participants in terms of ethics (see Manzo and Brightbill, Chapter 5 in this volume). More recent concerns have included what happens to participants once they return to potentially hostile environments beyond researchers' control (see Kesby *et al.*, Chapter 3 in this volume). However, we question whether PAR spaces themselves are always as 'safe' as researchers like to imagine. In 2000 Lee-Treweek and Linkogle edited an excellent collection of papers 'Danger in the field' which highlights the safety issues facing all social researchers; Box 6.1 summarises some of the key issues here.

PAR researchers are quick to celebrate the ordinariness of participants and the term sometimes takes on a rather sacred tone, suggesting benign, eager to be involved, perhaps disenfranchised persons. This can blind us, however, to the sometimes harsh and unpredictable realities of the vulnerable groups with whom PAR researchers often work and obscure the simple fact that some ordinary people and some everyday places can be dangerous. Take, for example, a researcher working with children on a project about air quality. The topic and participants may appear to be inherently safe but understanding the background of the children and making a serious risk assessment is still essential: while processes such as 'enhanced disclosure' in the UK protect child participants from abuse by researchers, children can themselves be violent and dangerous. Moreover, just because a research topic like air quality seems 'safe', we should not assume that discussion will not arouse memories that resonate dangerously with participants' personal biographies.

One can be seduced into a false sense of security when working with people for a period of time, but even after extensive interaction we may still know relatively little about particular participants' characters or what they are capable of. Participants and participant researchers are not required to go through the same

**Box 6.1 Potential dangers in the field**

- Physical (e.g. threat, abuse or sexual harassment from participants, researchers, or strangers)
- Emotional (e.g. resulting from actual or threatened violence or harassment, or from information disclosed or experienced during research interactions)
- Ethical (e.g. causing psychological or physical harm to others)
- Professional (e.g. positionality and/or role of the researchers)

Source: adapted from Lee-Treweek and Linkogle 2000

formalised vetting procedures as academic researchers (applications, curriculum vitae, references, 'disclosure' etc.) and so an element of vigilance and personal safety awareness should be maintained. At the same time as building up a rapport with our participant researchers, we must consider the safety issues implicit in the relationship.

## Issues of physical safety

In this section we consider some of the physical dangers PAR researchers might face. Using vignettes from our own experiences of fieldwork we illustrate the importance of taking risk assessment seriously and highlight how, with forethought, dangers might be avoided.

PAR often involves 'grassroots' work in 'everyday' but unfamiliar environments. Advocates take pride in conducting research in a variety of locations, including tenants' halls, people's homes, under trees, in religious buildings etc., rather than in sanitised spaces unfamiliar to participants (e.g. university campuses). However, our notions of working with and in 'communities' should not dull basic issues relating to safely accessing, working in and withdrawing from such environments (see Box 6.2 and Box 6.3). A consequence of flexibility is the necessity to consider the safety implications of working within each given space. Similarly, the common practice of working with groups rather than individuals should not reduce our awareness of the risk of personal assault (for us *and* participants). This could happen *within* an acrimonious group meeting, hence the need for 'conflict management' training, but could equally happen before or after a meeting.

### Box 6.2 Gemma's story

I led a feasibility study with a number of tenants' and residents' groups, exploring the potential for residential estate gardening clubs. Budgetary constraints meant I undertook this work alone. I agreed to visit one group at 6.30pm during their meeting at the local tenants' hall. I was given detailed directions on how to get there (including local bus routes). It was winter and already dark when I got off the bus. After a short walk I entered the housing estate looking for the hall. Moments after entering the estate I was aware that the route was not straightforward, the layout of the estate was confusing and I became disorientated. Initially I was concerned about being late for the meeting: annoyed at not giving a professional first impression. After more wandering my concerns changed to those of safety. Where was I heading? How could I get back onto the main street? Had I remembered to let my colleagues know where I was going? Did I have the phone number of my contact on me? Unfortunately my answers to these questions were not very reassuring.

Source: Gemma Moore

**Box 6.3 Mags' story**

A colleague and I once concluded a home based interview with the seemingly innocuous 'warm down question': 'How neighbourly is your neighbourhood?' The participant responded describing an occasion when a gang member threatened him with a knife. He claimed to have disarmed this person with a pen. Pointing the nib end of the pen he was holding at his jugular he suggested that a pen could be very dangerous if you knew what to do with it. I became concerned about the participant's state of mental health and about the risks of remaining in the flat, particularly when he asked 'has anyone in the room ever killed anyone, killed more than one person,' he continued up to ten and then raised his hand in the air. Still holding the pen he approached my colleague and pressed it against his chest. I was paralysed with fear given the participant's previous graphic descriptions, but knew I had to do something to bring the participant back to the present without further aggravating him. I identified a DIY project, still in progress on the floor, and asked questions about that. My colleague and I finally made our way to the door and left.

Source: Mags Adams

Gemma's experience (Box 6.2) highlights a number of safety issues (working alone or out of hours, working in unfamiliar locations). It also illustrates how researchers and participants can be put at risk by a decision to call a group meeting on a winter's night in a remote, poorly lit location that is difficult to find. At one level, it might be convenient for parents and workers to meet when children are asleep and the working day is over, but we have a responsibility to encourage participants to consider the safety implications of such an arrangement: for them and us.

Many of the methods used within PAR (diagramming, transect walks, role play etc.) may spin off in exciting directions during sessions. Like all good qualitative research they allow researchers to access people's life worlds, but also encourage participants to reflect on and analyse those life worlds. Ethically, but also in terms of safety, it is important to think through the possible ramifications of asking participants to ponder particular questions in particular ways and about the direction that a question, tool or process may take participants. Participatory research design can help avoid or manage some of these problems where collective memories and shared experiences and meanings can be identified as likely to trigger predictable responses and reactions. However, a review of methods is only effective up to a point: there was really no way that Mags could have anticipated that the question 'how neighbourly is your neighbourhood?' would produce such an extreme response from that one individual. There had been no adverse effects during 51 previous interviews, and no amount of piloting or participatory design could have predicted it. The risk of bringing up subject matter that triggers someone with (previously undisclosed) mental health, drug or relationship problems, and/or aggressive tendencies is difficult to determine particularly in the early stages of research or where researchers know little about participants' backgrounds or personal biography. Nevertheless, the key point we draw from Mags' experience is that the situation could have been managed more effectively. If the researchers had action plans covering a range of possible threatening scenarios, the research team could have withdrawn in a matter of minutes rather than the half hour that it took to extricate themselves from the situation.

## Issues of emotional safety

Issues of *physical* safety tend to dominate when researchers do actively think about safety before embarking on projects, but this is often to the detriment of considering *emotional* safety. The emotional distress often endured as a result of 'up close and personal' qualitative research that exposes us to accounts of the harshness of other people's lives is, yet again, little discussed in the literature. Experienced researchers perhaps see it as a 'hazard of the job' and very often we do too little to prepare those we train to expect and manage such distress. Here we focus on the more extreme and personally experienced forms of emotional trauma that can result from research.

Clearly, emotional distress may be an after-effect of physical trauma or witnessing a physical threat such as the incident described above, but in ethnographic and grassroots research it might also result from directly witnessing the poverty or suffering of others. Additionally, participants may recount something shocking and unexpected that the researcher must process and cope with at the time and attempt to rationalise in their subsequent analysis and representation (see Cahill and Torre, Chapter 23 in this volume). Being asked by participants to keep unexpected or particularly difficult confidences is another emotional burden that can cause researchers emotional stress, both 'in the field' and afterwards, depending on the nature of the confidences revealed. Shocking experiences may have a particularly intense impact on a researcher when the thing they hear or witness resonates with a traumatic experience in their own biography. Typical symptoms of this kind of emotional stress are insomnia, depression, headaches or flashbacks, and are no small matter. Researchers should therefore take their own past personal experiences into consideration when designing questions and choosing tools, locations, target groups and so on.

## Strategies and suggestions: embedding safety in research planning and developing a code of safe practice

So what should be done in advance of any research project to avoid unanticipated and unprompted physical and emotional risks? What strategies could have been used to extract the researchers promptly from the situations described above? The key to both of these questions is embedding safety in research planning and developing a code of safe practice. But whose responsibility is it and how should it be implemented?

Craig *et al.* (2000) maintain that safety at work is the responsibility of both employers and employees. In the context of social research, however, it may be hard to clarify how responsibilities should be shared. This issue has particular pertinence in PAR where participant researchers are usually not engaged via the kind of contractual arrangements that cover 'principal investigators' and 'research assistants' employed in conventional research. Furthermore, PAR researchers often train and engage others to undertake research with groups and in environments that they otherwise could not reach. These people and places might be dangerous, or might become so for the participant researcher precisely because of the activities full-time researchers have asked them to carry out. Consider the researcher who gives disposable cameras to street children and asks them to photograph aspects of their everyday lives. Some activities might be illegal, and photographic evidence might incriminate participant researchers or those photographed. Full-time outside researchers (or employed academics) need to be aware that engaging participants as 'researchers' may change their status in the eyes of other community members and lead them to face hostility as 'outsiders'. Thus the informality and flexibility of our arrangements with participant researchers, and our efforts to share research design and blur roles and responsibilities in our projects, should not lead us to abrogate our responsibilities for our participants' physical and emotional safety. Full-time researchers still have a responsibility to oversee safety procedures within the collaborative process and to intervene to ensure the safety of *all* those involved in a project (see Box 6.4).

**Box 6.4 Taking responsibility for safety**

- Universities or employers have a duty to provide researchers with safety training.
- Ethics Committees should ensure that safety assessment and appropriate training is established before projects receive ethical clearance.
- Principle Investigators (PIs) bear an ethical responsibility for the safety of participants, researchers and participant researchers recruited into research processes they instigate.
- PIs should conduct a thorough risk assessment of the project. This should include assessments of (a) lone working (b) research locations and times and access (c) threats posed to *and by* participants/participant researchers (d) methodology and tools (e.g. questionnaires, *aides-mémoires*, diagrams, focus group topics, drama) to identify any potentially inflammatory questions or situations.
- PIs should develop 'actions on' that allow researchers to respond quickly and safely to potentially dangerous situations.
- Universities or employers should provide or facilitate counselling for employees who have encountered particularly traumatic experiences as a result of their work.

Source: Authors' own analysis

In Box 6.5 we outline some recommendations for a code of safe practice. Learning distraction techniques, developing code words to alert colleagues to a danger and the trust necessary to act on a colleague's fearful feelings without question, are all central to maintaining a safe research environment. It is too much to expect that risk situations will never arise, but it is possible to be prepared by

**Box 6.5 Key procedures for personal and/or personnel safety**

- Make 'safety' a subject for discussion with participants when discussing 'ground rules' and make safety review part of PAR design.
- Conduct a participatory risk assessment of the project with participant researchers.
- Provide safety training for all members of your research team. Include role play as people remember how to react in a situation if they have enacted it previously.
- Organise an emergency contact person who knows: (a) who is in the research team, (b) the locations and/or addresses of research sites, (c) what you will be doing and (d) when you are expected back and/or to make contact. Ideally you should call the contact person before every research activity and tell them when it will finish (they might set an alarm clock). If research over-runs, call to explain (let participants know this is part of your safety procedure). Phone your contact when you finish. If you do not call, the contact person must phone you (so leave your phone on). If they cannot reach you they should contact the authorities. Consider adding mobile phones and/or top-up cards to grant applications for this purpose.
- It is possible to use a commercially available computerised telephone check-in system (costly but effective). Encourage your institution or employer to invest in such a system.
- Establish a series of 'actions on' covering a range of scenarios that enable full-time and participants researchers to respond to hazardous situations in a pre-arranged fashion.
- Establish safety code words (such as: 'we need the purple file') that signal to co-researchers that all is not well and trigger a pre-arranged response (variously this might be: change the topic, stop the research, withdraw from the research context, depending on the code word used).
- Review safety issues and procedures regularly. This is particularly important in PAR because new themes, questions, tools, and approaches may be created within the project, the risks of which may not have been anticipated before research began.

Source: Authors' own analysis

having strategies available. Ironically, if no incidents occur there is a danger of complacency and lapsed attention and commitment. Thus safety procedures need to be reviewed on a regular basis ideally through debriefing after key stages of the research process, both within the research team (academics and participant researchers) and within the academy (academics and ethics review processes).

## Final words

The objective of this chapter has not been to completely resolve the tensions within research ethics as they relate to researcher safety. Rather we have tried to highlight the importance of considering the risks that researchers may face in the field alongside their ethical responsibilities to participants.

With PAR the blurring of the boundary between who counts as the researcher and who counts as 'the researched' adds a further layer of complication to the issues we have been discussing. While we advocate institutional provision and support of safety procedures, these provisions are often not accessible to participant researchers who are not formal employees. Consequently, PAR researchers within the academy have even more to contend with than other social researchers as they need to consider not only the safety of themselves and their colleagues but also the safety of any participant researchers. We therefore appeal for a more reflexive evaluation of the hazards and risks associated with methodologies and tools that have been specifically designed to be inclusive and participatory.

Finally it is important to stress that the majority of PAR researchers have positive research experiences. Nevertheless, through the implementation of excellent safety procedures we feel all researchers will be safer without compromising the quality of their research.

## Notes

1 However, for useful examples, see Hughes 2004; Paterson *et al.* 1999; and the journal *Qualitative Research* (e.g. Fincham 2006; Nilan 2002; Sampson 2004; Sampson and Thomas 2003).

2 Our experience is primarily with qualitative methods (interviews, questionnaires, photo surveys, soundwalks and oral histories); however, Gemma worked on several participatory projects while employed by a voluntary organisation.

# Part II

# Action

# 7 Environment and development

## (Re)connecting community and commons in New England fisheries, USA

*Kevin St. Martin and Madeleine Hall-Arber*

### Introduction: participation and its effects

There is an extensive literature on participation in environment and development that has made clear the range of issues one must consider when engaging in participatory development projects to achieve particular development and/or conservation goals. There is, however, less review and commentary on the work that participation does, or might do, relative to the discourse of development itself. That is, it is well known that participation alters and re-thinks methods of development implementation, but how is it constitutive of alternatives to standard forms of development and/or conservation? Beyond participation itself, what are its effects?

Our intention in this chapter is to point to what we think are the effects of participation, the work it does relative to our imaginaries of environment/development, and to link those effects to our hopes for sustainable and progressive economies and environments. Throughout, we reference the case of fisheries, a contemporary site of struggle over the form of development that will best conserve fish stocks and habitats. We describe the effects of a recent Participatory Action Research (PAR) project in New England, USA that was designed to disrupt and replace the dominant image of fishermen's[1] behaviours and environmental practices. PAR projects can, we conclude, effect change by producing alternative ontological and discursive foundations vital for imagining sustainable environmental/economic futures.

### Constituting alternative economic and environmental foundations

Critics of participatory development suggest that it is implicated in the neoliberal development project insofar as it enrols individuals at the local level into processes and relations of economy aligned with that project (for example, Cooke and Kothari 2001a). In addition, they point to the problems of assuming pre-existing communities and overwriting the complexity and unevenness of local society (see Kesby *et al.*, Chapter 3, and Pain *et al.*, Chapter 4 in this volume). While we must carefully consider these critiques, we must not allow participation to be subsumed into a discourse of a single global economic order with its ability to penetrate all localities (cf. Gibson-Graham 2003). Indeed, participation has long been posited as a method for discovering alternative paths and possibilities that emerge directly from localities and communities themselves. If we look not just at how participation is aligned with neoliberal development but also at the many ways it disrupts and contradicts such strategies by positing alternative ontological and discursive foundations for environment/economy, we might see how participation can be transformative rather than tyrannical (Cameron and Gibson 2005a; Hickey and Mohan 2004).

To illustrate, we briefly turn to the example of participatory mapping, nearly ubiquitous in participatory and integrated environment/development schemes (cf. Sanderson *et al.*, Chapter 15 in this volume). Participatory mapping has emerged across a number of sites as a means by which local landscapes and resources can be inventoried, their local use documented, and sovereignty over them claimed (for example, see the special issue on Geomatics, *Cultural Survival Quarterly* 1995). From rough sketches of agricultural practices in Rapid Rural Appraisal settings to detailed digital mapping of territories, land cover, and traditional use patterns with common lands, participatory mapping has thoroughly infiltrated participatory interventions focused on environment/development. While many mapping projects smack of information extraction and/or the use of local labour for spatial land use data collection that would otherwise be expensive or impossible to obtain, others are resolutely aimed at 'counter-mapping' community territories and resource commons relative to the threat of their appropriation or exploitation by an external and often multinational capital (cf. Herlihy and Knapp 2003; Peluso 1995).

While participatory and community-based mapping might be criticised for ignoring the micro-politics of access to and use of the areas and resources being mapped (for example, Hodgson and Schroeder 2002), here we are interested in how such mappings generate alternative ontological and discursive foundations for re-thinking local economies and environments (also see Sanderson *et al.*, Chapter 15 in this volume). That is, counter-mapping proceeds by first positing a locale or territory within which is embedded a community and its resource utilisation practices. The community is then engaged directly in the documentation of space and resources (e.g. their quality, location and seasonality) and, through cartographic representation, an inhabitation and productive use of that environment by the community is established. Knowledge of the environment and its use is inscribed as a community-based practice within a common space. The social relations and practices of resource use, as well as the community-based process of mapping them, become processes constitutive of community and commons. Participatory community mapping creates an alternative discourse of space and its inhabitation relative to state-sponsored boundings, privatisations, categorisations, and inscriptions (cf. Stocks 2003).

To address critiques of participatory development that accuse it of reifying homogenous communities and furthering the neoliberal economic agenda, we might, building on Offen's work (2003), recast community and commons as effects of participation (and participatory mapping) such that our research could then document their emerging character *vis-à-vis* participation, foreground the creative possibilities of community and commons, and build upon their potentials relative to alternative (and sustainable) economies and environments. Furthermore, if we understand community and commons to be processes and social relations rather than bounded and discrete entities, our participatory research would not so much search for those discrete entities but facilitate processes constitutive of community and commons, as is the forte of Participatory Action Research.

## Disrupting the developed economy and reclaiming the environment

In the minority world, there has been more limited use of participatory methods to address questions of the environment and development (particularly, their nexus). While there is considerable growth in participatory approaches, that growth is primarily concerned with what we see as responses to development rather than development *per se*. That is, participatory approaches are not so much about envisioning economic alternatives or economic diversity via participation (although see Cameron and Gibson 2005a) but are about responding to an existing economy and its practices of environmental degradation, toxification, or undermining of local economic and community well-being. Examples of such responses to the effects of development include cases of environmental injustice due to noxious industrial output, community health movements where disease clusters are tied to industrial practices or industrial products, struggles over the development of remaining open space or struggles over wilderness use, and the problems of resource scarcity and allocation between increasing numbers of stakeholders.

While such projects lack a development agenda insofar as they are not about producing an economy or facilitating new economic and productive practices (even locally), they effectively point to the possibilities of participation as an ethical practice of community formation and solidarity relative to environmental injustice, or degradation due to development (see also Lawrence 2006). Indeed, the 'success' of such projects is often measured in such terms rather than in terms of new production capacities or economic growth as in the majority world. We are interested, however, not only in projects in the minority world that constitute community and a common interest as forms of resistance and solidarity, but also in how these emergent processes might foster alternative economic practices and developments as they are free to do in the majority world (St. Martin 2005). How might participatory projects create new community-based claims to space and environment even in the midst of a hegemonic economic ideology of private property and individual interests? How might sites 'within' capitalism, sites where community and commons have been assumed to be long gone, also be sites of alternative economic and environmental possibilities? (St. Martin 2006).

## (Re)producing a fisheries commons

A US federal programme designed to promote 'cooperative research' between fishermen and scientists funded the project we present here.[2] The programme was a response to the industrial/environmental crisis in fisheries in the 1990s and, amongst other things, it served indirectly to funnel funds to struggling fishing communities (Hartley and Robertson 2006). While most projects were concerned with designing and testing new forms of fishing gear or testing scientific hypotheses based on fishermen's knowledge of local fish stock, ours would be an action-oriented and participatory research project that worked from the concrete socio-economic and environmental experiences of fishermen toward new understandings and, literally, images of fishing communities and their territories. Working with a team of 'community researchers' we developed a mapping and interview-based protocol that they would deploy in a range of New England communities that fish within the Gulf of Maine.

The protocol sought to disrupt the dominant image of commercial fishing as essentially the practice of individuals competing within a space open to all and across which all could freely roam. We were interested in creating maps and narratives that would locate fishermen as embedded within a variety of community processes and would link their community and fishing practices to common fishing grounds and specific fisheries habitats. Our work, we hoped, would be a 'counter-mapping' relative to the dominant discourse of fisheries which marginalises communities and their ability to participate in fisheries science/management and undermines the maintenance of a fisheries commons as an ethical practice of sharing.

At an ontological level, the project is an empirical collecting of information that is missing or overlooked by standard forms of fisheries science and management. In this sense (and this is how we represented the project to the funding agency as well as the prospective participants), our project would inquire as to the existence of fishing communities within the region, it would investigate the nature of those communities, it would document the range of environmental knowledge held by community members, and it would do so via mapping methods such that communities and fishermen's knowledge would be explicitly linked to locations at sea.

At the level of discourse, the project is an explicit attempt to enlarge the field of possibility in natural resource management as well as local resource-based economies (Gibson-Graham 2005). That is, the project attempted to constitute a literal space into which participants might project themselves as community members, in relation with other fishermen, as embedded within a locale, and as knowledgeable subjects (cf. Community Economies Collective 2001). We see the project as giving fishermen (and here we mean both our community researchers and their interviewees) a forum for imagining themselves in ways that differ from their depiction/subjection by the dominant scientific and managerial discourse of fisheries in the region.

## The project process

The first stage of the project involved the production of a series of maps depicting the fishing locations of fishermen from particular ports. These maps were produced solely from federally mandated vessel log book, data which record fishing trip locations on a trip by trip basis. Using Geographical Information Systems (GIS) techniques, the locational data were aggregated by fishermen peer group (i.e. by port and gear type) and superimposed onto standard nautical charts. While these maps uniquely represented the territories of fishing communities, they could only act as catalysts for community and an alternative community-based management of fisheries if they were integrated into a PAR protocol.

To develop such a protocol, we hosted a workshop which was attended by fifteen potential community researchers from several communities in New England. At the workshop we represented the project in the first sense noted above. We presented the maps we had made and discussed how they might be useful to fisheries science or management, as well as fishing communities interested to advocate for continued access to resources and participation in management. Finally, we suggested that the maps were only a first step and that to be legitimate and useful they would need to be vetted, amended and given meaning by fishing communities themselves.

The project was intended, however, to do more (see above) and at the workshop we insisted that the project should not be an assessment of what was disappearing but an exploration into how fishing communities, their territories and knowledge of the environment could be foundations for the future. Initially, this proved to be uninteresting to the workshop participants given what we soon recognised as a lament for a 'way of life' (culture, community, family businesses, etc.) that is in decline throughout the region. The very real financial and cultural crises that are accompanying the crisis in resources are overwhelming. To develop any sense of community possibility in the face of this lament we turned to the maps and their novel depiction of communities and territories within what is usually a space devoid of either (Plate 7.1).

We relied upon fishermen's intimacy with the nautical charts, which were the backdrop for our data outlining community fishing territories. Such charts are part of their everyday lives and act as the discursive frame for their community interactions, sense of place and knowledge of the environment. While the technology and use of nautical charts were very familiar to our participants, our superimposition of 'their' fishing grounds (derived from 'their' log book entries submitted to the government) onto these standard charts was unexpected and created an almost visceral reaction. Fishermen were seeing for the first time an outline of the territories and locations of greatest importance to them and their 'communities'.

Unlike most participatory mapping projects that start from essentially blank maps and ask participants to draw their territories of use and practice, ours used GIS and mapping techniques to disrupt standard images of fishing space as devoid of human habitation, to present the space of fishing as a space of communities, and to do so within a cartographic frame/language that fishermen used every day. While the project participants all debated the accuracy of the maps, the boundaries of communities, and the eventual use of the maps, they were all clearly drawn to the maps as novel and surprising representations of their working lives.

> New forms of subjectivity emerge through unexpected shifts in the visceral and affective registers that free embodied practices from their usual

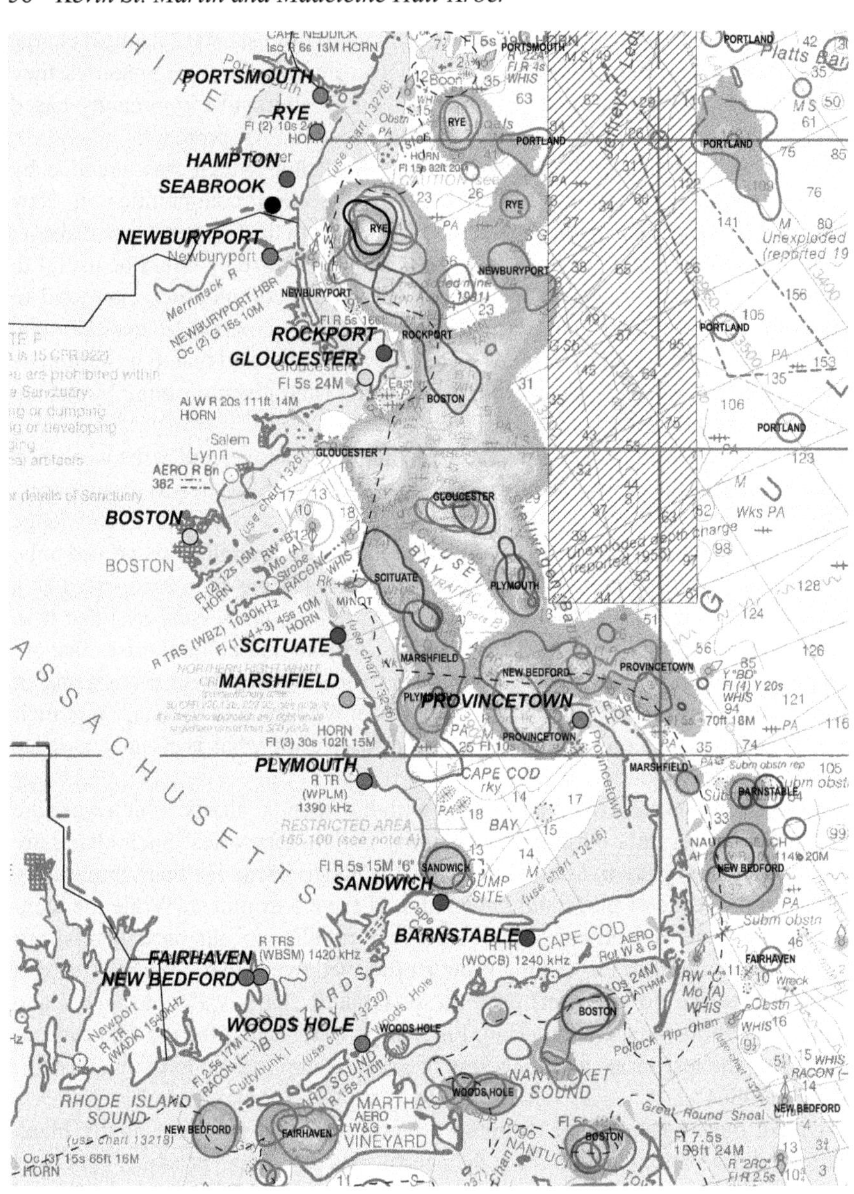


*Plate 7.1* Outlines of community territories. An extract from a chart with colour-coded outlines of community territories superimposed upon a NOAA nautical chart. Areas outlined represent primary fishing grounds by port (Source: St. Martin 2005).

> sedimented patterns, creating opportunities to act on other possibilities for being [...] We are, however, interested in harnessing the creativity of more everyday events that might inspire previously unknown possibilities and increase a willingness to explore different ways of being in the world.
>
> Cameron and Gibson 2005a: 320

The maps convinced the workshop attendees to pursue the project with us and to work as community researchers. Which maps, their look, and how they would be integrated into an interview schedule was hammered out at the workshop. The community researchers then took the revised maps and interview schedules back to their communities (prominent fishing ports in Massachusetts, New Hampshire, and Maine) where they interviewed fishermen who worked in a variety of fisheries, from different ports, and with a wide range of fishing gear and vessel types.

The interviews began by asking fishermen a series of questions meant to prompt them to locate themselves within a peer group/fishing community and to assess its community or cooperative nature. The bulk of the interviews consisted of mapping exercises that moved from the region-wide scale of the Gulf of Maine down to the locations most closely associated with the interviewees' communities. For each of three maps, those interviewed were asked to assess and amend the map, to discuss its accuracy, and to describe how they (as representatives of a peer group) inhabited the spaces depicted, who else works in or inhabits those spaces, what are the relationships between proximate fishermen, and what they know about the environmental conditions and processes of those spaces.

## Project outcomes

While the project outcomes are varied and still emerging, we are heartened by recent responses to our initial summaries and public presentations of the project. Indeed, we have been surprised by the degree to which maps of the 'social landscape' of fishing are intriguing and occasionally disruptive across a number of sites. As the product of both federal data and fishermen's participation, the maps are at once scientific and widely legitimate representations of what had been primarily 'anecdotal' information. Aggregated, formally presented, and vetted by fishermen themselves, data representing community territories are becoming weighty icons of the 'human dimension' of fisheries (cf. Jensen and Richardson 2003).

At the ontological level, this participatory project has produced data – maps and narratives of community processes and commons practices – that had not previously been collected. To use a GIS metaphor, the project has produced a layer of information that had until now been missing from fisheries science and management. As fisheries science and management is indeed turning toward more spatial approaches, particularly in the guise of ecosystems-based management, there is a real potential for these data to act as an actual layer in the analyses and management of fisheries. Fishing communities and their territories, once relegated to the majority world may, after all, be significant contributors to the current regime of

**Box 7.1**

*Community researcher*: I know these people, the fishing, but the [interview] questions allowed me to go deeper into their feelings. The thing that was amazing to me, fishermen care for each other, because of what is happening to them. I [interviewed fishermen from] two different gear types, I did draggers and I did gillnetters. The thing that really startled me: you hear that they really don't like each other because of the gear conflict [...] but what was surprising was the worry of the gillnetters about what was going to happen to the draggers under these regulations, and the amount of cooperation, that is bringing them together. They are sharing their [allocated fishing] days, and if a guy doesn't have money, they are letting them have the days [for a] minimal amount of money so they can stay viable. [A] new bond is being created in the world of fishermen.

Source: Community researcher 2005

fisheries resource management. In addition, their revelation and representation makes more likely their participation in decision making and collaborative fisheries science.

At the level of discourse and its capacity to define and inscribe both fishermen and space, we note a number of instances where the singular discourse of fishermen as individuals competing within a homogenous and open access commons was disrupted by project participants. This was clearly evident at our final workshop where community researchers expressed an altered understanding of community and its possibilities (see Box 7.1).

In addition, the project's premise and outcomes have informed the recent development of the 'Area Management Coalition' – a network of fishermen, academics, and sustainability advocates – who are proposing an area-based and community-centred form of fisheries management in the Gulf of Maine. Founding members of the coalition were also participants in our project and they remain interested in collaboration. It would seem that 'fishermen' are emerging as other than subjects of the dominant discourse; they are re-positioning themselves as community members, as participants in common environments, and as visionaries of alternative resource management schemes.

## Conclusion

While the turn to participation by development and international environmental organisations can be criticised as yet another way to enrol local communities and individuals into neoliberal economic logics, we have focused on the positive effects of participation that continue to work as critical counter-measures to standard forms of development (see also Kesby *et al.*, Chapter 3 in this volume). In particular, the insistence on participation as a necessary element of economic and environmental initiatives serves to shift the scale at which we imagine such interventions, insofar as participation presumes a community of participants and a locality within which that community resides or works. Participation foregrounds community and commons and suggests that they might be foundations for economic development or environmental management. In addition, participation can provide the conditions for transformations of subjectivity and spatiality; for disruptions to the fixing of both by dominant environment/development models. In our work in fisheries, we rely upon Participatory Action Research to enable a cartographic re-visualisation of space and community inhabitation, which shifts identifications and opens up the environment and local development to new possibilities.

## Notes

1 The term 'fisherman' will be used throughout to refer to persons directly involved in fish harvesting. While not gender neutral, it is the preferred term by both men and women who work as harvesters of fish in New England.

2 'An atlas-based audit of fishing territories, local knowledge, and community participation in fisheries science and management' was funded by NOAA via the Northeast Consortium (#01-840). Principal investigators were Kevin St. Martin, Rutgers University and Madeleine Hall-Arber, MIT Sea Grant.

# 8 Working towards and beyond collaborative resource management

## Parks, people, and participation in the Peruvian Amazon

*Michael C. Gavin, Alaka Wali and Miguel Vasquez*

### Introduction

We believe participatory approaches to conservation action can help communities and interested external agents overcome some of the barriers to successful collaborative or community-based resource management. In this chapter we outline our work with communities neighbouring the Cordillera Azul National Park in Peru. The participatory work we have been a part of in the region has sought to gather vital data about resource use to inform the Park's management plan, as well as to empower communities and build local capacity to engage in collaborative resource management.

### Background

Perhaps the most vigorously fought debate in the field of conservation for over a decade has been between advocates of strictly protected areas, and proponents of community-based conservation. The protectionist approach, sometimes called fortress conservation, has been a dominant paradigm since the first national parks were established in the United States in the mid-nineteenth century. Opponents of protectionist conservation strategies have pointed to ethical and logistical problems inherent in this approach (cf. Wilshusen *et al.* 2002), including issues related to indigenous sovereignty, lack of funding for enforcement in protected areas, and the value of local traditional ecological knowledge. Critics of fortress conservation argue that the question is not if local people should be involved, but rather how they should participate.

Alternatives to protectionist conservation allow varying degrees of local participation in natural resource management. Similar to participation continuums (see Kindon *et al.*, Chapter 2 in this volume), a broad spectrum of possibilities exists (see Figure 8.1) (da Silva 2004; Sen and Nielsen 1996). Protectionist forms of conservation tend to fall toward the lower end of such continuums because they are externally driven. These less participatory means of management may only

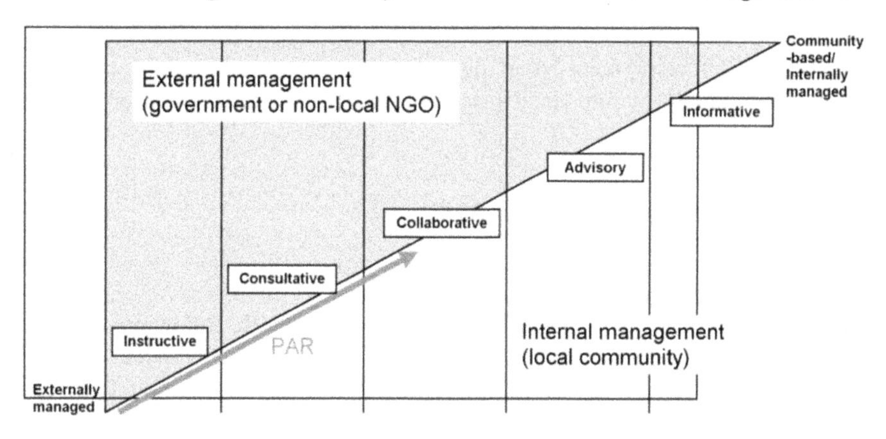


*Figure 8.1* Resource management strategies and the role of local communities (adapted from Sen and Nielsen 1996)

involve external agents instructing communities about management rules, or occasionally consulting with local people about desired outcomes before taking unilateral action. More participatory management models (e.g. CAMPFIRE in Zimbabwe) would fall in the upper part of any continuum. Here more power is exercised by local communities, which may seek advice from outside parties, or simply inform external organisations of their management actions. In the middle of this range are collaborative arrangements in which communities and external agencies share responsibilities and benefits. Employing a Participatory Action Research (PAR) approach can encourage more local-level participation and empowerment, effectively driving the relationship between external organisations and local communities towards more collaborative arrangements or beyond to community-based models.

For more collaborative or community-based conservation to achieve success, external agents must be willing to cede power, communities often need to build critical capacities, and the social and environmental context must be appropriate. External players who are unwilling to recognise the rights of local communities over land and the use and management of resources stand in the way of collaborative or community-based conservation (Ostrom 1990). Similarly, weak institutions at all levels that fail to promote transparent management and fight corruption often lead to unequal participation and inequitable distribution of the benefits accrued from natural resource management (Brechin *et al.* 2002). Local institutions, including relevant social norms, are also critical for effective management; and the strength of these institutions often depends on capable leadership and mutual trust built from community social cohesion (Ostrom 1990; da Silva 2004). In addition, programmes for which stakeholders and resources are not clearly defined, and/or where local communities do not maintain a working knowledge of, and reliance upon the resources in question, are more likely to fail (Adams and Hulme 2001). Without resource dependence, communities may lack a vested interest in sustainable natural resource management. Finally, the long-term viability of collaborative or community-based projects can be compromised by political, economic, social or environmental instability at local, national, regional or international scales.

## Conservation in the Cordillera Azul

### *Brief history*

The Cordillera Azul National Park was gazetted in 2001, following a Rapid Biological Inventory conducted by The Field Museum, together with local collaborators (Alverson *et al.* 2001). Located in Northern Peru, between the Huallaga and the Ucayali Rivers, the park is over 1.3 million hectares (roughly the size of the US state of Connecticut) with a buffer zone of over 2 million hectares (approximately the size of Wales) (see Figure 8.2).

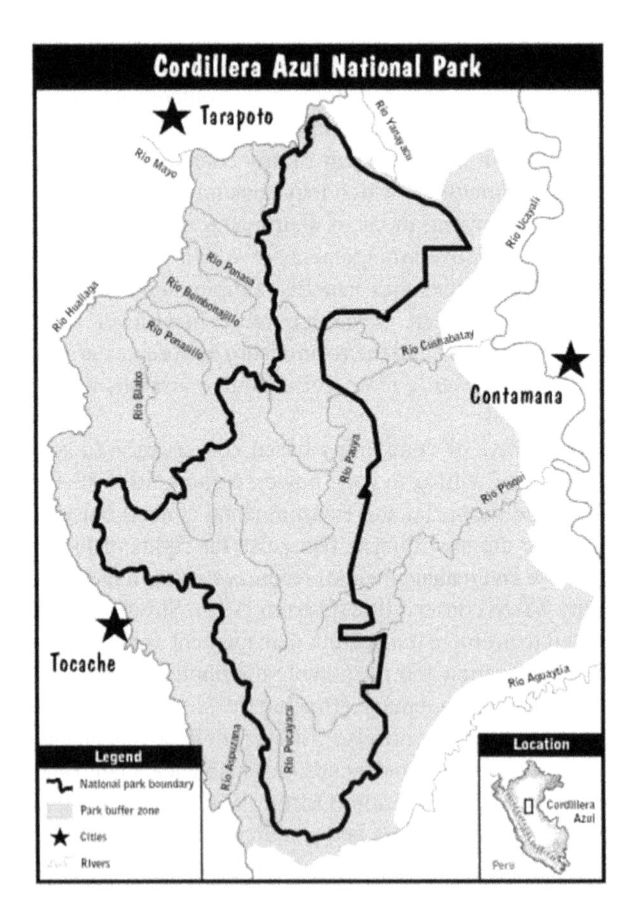


*Figure 8.2* Map of the Cordillera Azul National Park, Peru

Ringed by sheer rock escarpments, the park contains mountain crests and ridges, broad lowland valleys, high-altitude lakes and marshlands. The dramatic landscape is home to an extraordinary diversity of plants and animals. Although no people live within the park's boundaries (early negotiations relocated the few families that had pastures or fields within Park boundaries), the western buffer zone, in the Huallaga River Valley, is more densely populated, containing over 100,000 inhabitants (of which approximately 25,000 live within direct access to the park). The Ucayali River Valley, in the eastern buffer zone, is sparsely populated by largely indigenous communities (Shipibo, Cacataibo and Piro).

The Non-Governmental Organisation (NGO)[1] awarded the contract to develop a management plan engaged local populations in the plan's design and implementation. The NGO's tactics deviated significantly from previous strategies used in Peru, where eight previously established parks were managed largely with protectionist policies.

### *Role of participatory methods*

Several participatory methods were employed during the development of the park's management plan. We highlight two methods with which we were personally involved: a participatory ethnobiological resource use study, and a resource use and capacities mapping exercise.

#### *Ethnobiological resource use study*

Michael, Miguel and two other Peruvian university student assistants began the ethnobiological study in October 2001 in the three communities closest to the park's northwestern boundary. At the beginning of the research many community members had no idea the park existed and others were immediately concerned

**Box 8.1 In the beginning**

In the beginning, the park was very controversial with some communities. There were communities that did not accept the park's creation because they relied upon periodic visits to hunting and fishing grounds for both subsistence and for the sale of dried meat, skins, and live pets, and for illegal logging. There was also the issue of land trafficking, with those involved taking a close and personal interest in the extent of the park's boundaries, as this affected their ability to increase land holdings for future dividing and selling. The participatory process was difficult at the beginning, because those working with the communities had to mediate the different agendas of community members, including those with personal interests or external agents who were illegally exploiting natural resources.

Source: Miguel Vasquez

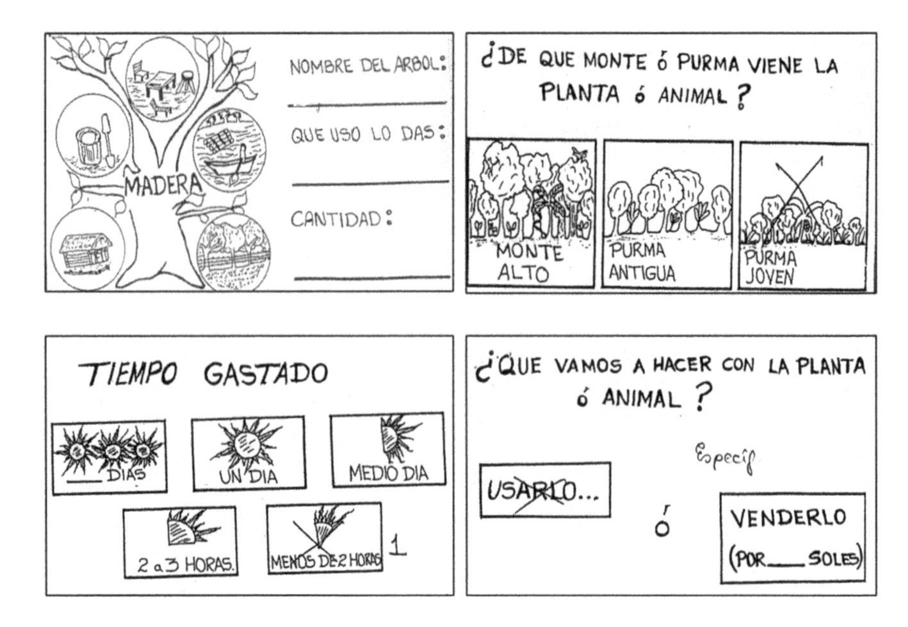


*Figure 8.3* Example of data sheets used in participatory ethnobiological survey

about the potential impact the park would have on their lives (see Box 8.1). Many participating families were interested in the research as a means of both improving their own land management practices and as a way to demonstrate their resource needs for any future negotiation with park officials regarding access to resources inside the protected area.

Michael developed a participatory ethnobiological method in which participating families recorded on a daily basis all flora and fauna they collected over a six month period. Use categories were employed to help families remember all resources collected. The categories (e.g. food, medicine, etc.) were defined at the beginning of the data collection process by local families using a combination of free lists, pile sorts, and hierarchical clustering. This process of defining the use categories was based on *in situ* culturally sensitive perceptions of forest resources, in contrast to many other ethnobiological studies, which employ *a priori* use category designations.

Families used paper data sheets (designed by Michael and Miguel in consultation with local families, see Figure 8.3) to record the name of species collected and the quantity (in locally appropriate units; for example, bundles of firewood), the forest type where the good was gathered (young secondary, old secondary, or older forests), the time spent collecting the good, and whether the good was sold or used in the household. Data collected over six months were used to examine the intensity of use of different species and different forest types, and to compare use patterns to estimates of sustainable use levels (Gavin 2004; 2007).

Results indicated that families were collecting 324 different species of plants and animals (Gavin and Anderson 2005). The data also highlighted the importance of open-access land, including forests within the park, as a critical source of resources (Gavin 2004). In addition, several species were being utilised at unsustainable levels, and management intervention would be required both to ensure long-term viability of the species and to secure the communities' resource base (Gavin 2007). The results suggested that intervention needed to include mechanisms to control immigration into the region and to provide alternative protein sources to meet local dietary needs. Finally, and perhaps most importantly in terms of the development of the park's participatory management approach, the ethnobiological study demonstrated the ability of local people to monitor their own resource use.

### *Resource use and capacities mapping exercise*

In 2002, to insure an inclusive process for the design and implementation of the master plan, the park team[2] developed another unique approach to participation, the *Mapeo de Usos y Fortalezas* (MUF – Resource Use and Capacities Mapping). The MUF incorporated elements of participatory social asset mapping (cf. Kretzmann and McKnight 1993) with resource use mapping (cf. Chapin and Threlkeld 2001; Herlihy and Knapp 2003) to elicit useful information for designing strategies to engage communities in park conservation efforts. The key elements of the MUF were as follows:

- Trained facilitators (see Box 8.2) from fifty-three communities who were responsible for collecting information. These facilitators participated in a four-day training workshop on conducting focus groups, surveys and interviews, creating resource use sketch maps, and reading satellite images.
- Facilitators received support from extension agents and anthropologists who helped analyse data.
- The resulting 'assets' database allowed extension agents to implement intervention activities based on a knowledge of communities' social strengths and capacities, existing resource use strategies and local environmental knowledge.

**Box 8.2 The importance of community facilitators**

I believe that the MUF was perhaps the most important activity we undertook in the communities, especially because it involved building the capacity of a democratically elected community member. This community member was called the Facilitator. This person served as a bridge or mediator for the park's work with the community. The Facilitator also stood out as the part of the process in which community participation became a part of decision making, and the Facilitator became the critical link between the park team and the communities.

Source: Miguel Vasquez

- The commitment to sharing information with local communities to provide leaders and community members with a sound basis for decision making and awareness of the park's significance.

The MUF was different from conventional Rapid Rural Appraisal or Participatory Appraisal methods (Kindon *et al.*, Chapter 2 in this volume) in several ways:

1. It took place over a longer time-frame (two months for data collection, six months for database creation) allowing for deeper participation and richer information that went beyond the views of community leaders and engaged diverse community members.
2. It focused on community social strengths and capacities, and did not rely on the traditional 'strengths, weaknesses, opportunities and threats' framework, which often does not provide sufficient understanding of intrinsic capacities and attitudes.
3. It provided community residents with satellite images, maps and other materials on the park and its buffer zone, thus giving direct access to information.

Initially the park team was uncertain how much valid data could be collected by community facilitators, given they had only the equivalent of a rudimentary high school education. However, by the end, some facilitators had filled entire notebooks with information, drawn detailed resource use maps, created visual elements that expressed communal identity (see Plate 8.1), and even written poems about the park. Ultimately, the database was linked to a Geographic Information System (GIS) programme, and the resource use maps, as well as the spatially explicit social asset data were geo-referenced. Much of the MUF data were included in the park's management plan.

The MUF, coming at the beginning of the management plan process, set the standard for all subsequent participatory activities. For example, the whole extension programme (focused on engaging communities to reinforce or adopt conservation-compatible livelihood strategies) is based on community participation, from the design of specific programmes to their implementation. The park protection effort also involves community members in the selection of local park guards, and the organisation of a volunteer guard corps.

### *Current status of park management*

A participatory approach has continued to characterise the park team's strategy (see Box 8.3). Building on the MUF results, mapping exercises are leading to consolidation of zoning into formal land use plans, which will be recognised by the regional government, giving communities a measure of land security and a mechanism to control migration and intensive resource exploitation. In this way, the MUF has been useful for empowering local communities by providing them a voice in the process of land management and by beginning the process of securing land rights. The park's management plan (INRENA 2006) calls for formal recognition of the partnership

*Plate 8.1* Example of a community shield produced as part of the MUF (Credit: CIMA-Cordillera Azul, translation by M. Gavin 2002)

**English translation**

1 In the first space, a male *paucar*, because we consider it to be a very intelligent bird that is a great imitator of domesticiated animals with its song; it is a very entertaining bird.

2 In the second space, a tree that represents the natural forests that we have, in this case it is a *manchinga* tree that is a medicinal plant. Its buttressess are of use in the construction of canoe paddles, also in artisal work (e.g. plates, buttons, and others of domestic use).

3 In the third space, the mountain or Cordillera Azul, with the rising sun, that signifies hope for a better future, and we are inspired looking at it from a distance, an unknown world, anxious to become acquainted with it.

**Box 8.3 Community participation**

Community participation is now much more than it was in the past. Communities not only contribute work to the park, but they also participate in decisions. Now, when activities are promoted in the communities, the communities can assume more responsibility for their own development, beginning with deciding which projects should be undertaken and leading to the mobilising of resources and the organisation of activities. The objective of promoting community participation is to establish that the decisions which affect communities should be made by all members of those communities, and not just a few or by external agents. However, the current situation remains divided. I believe this is due to the constant immigration into the communities where we are working. And we must also not forget that there were always people present in the area that were never in agreement with the conservation activities of the park, due to their personal interest in the trafficking of land, illegal logging, and commercial hunting.

Source: Miguel Vasquez

between park officials and local communities through *Acuerdos Azules* (Blue Agreements), which commit the park and the communities to certain actions. The park has also adopted a participatory adaptive management model (see Figure 8.4), in which communities are actively involved in creating resource use rules and in monitoring and managing resource use, using techniques first developed in the ethnobiological study outlined above (INRENA 2006).

## Limitations

The park's management plan is clear evidence that the efforts of communities, researchers, NGOs and government (through its agency INRENA) has led to remarkable progress in moving towards more collaborative management. However, what is on paper can oversimplify reality on the ground. Three challenges to more participatory conservation that remain in the Cordillera Azul have also been recognised as major barriers to more participatory conservation approaches in many locations. These are:

- Government's unwillingness to cede control over resources (e.g. Ostrom 1990);
- Funding constraints (see Dearden *et al.* 2005); and
- Exogenous pressures on resources (Sunderlin *et al.* 2005).

One critical step in developing the Cordillera Azul's management plan was INRENA's realisation that government resources were insufficient to cover another new park. New legislation permitted private organisations to be involved in

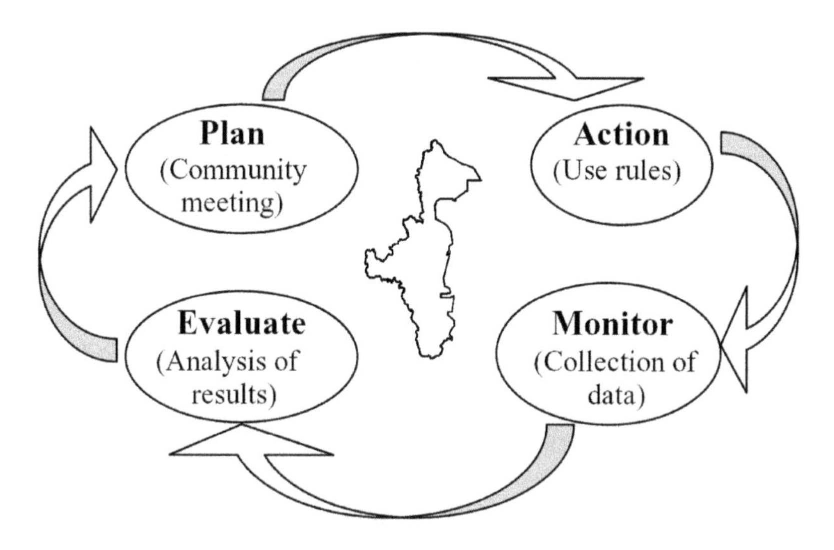


*Figure 8.4* Participatory adaptive management plan for the Cordillera Azul National Park, Peru

national park management, and the Cordillera Azul was to be the first trial of this system. The NGOs involved had access to international funding to help support plans for a more participatory strategy. However, the relationship between CIMA, as contracted park 'manager' and INRENA, as the government agency with authority over the park, has sometimes been tense because lines of control are not clearly established. At times, INRENA's more bureaucratic (and less participatory) methods have taken precedence. It was difficult to create a more two-way information flow between community residents and professional staff because of constraints on the number of staff and time allocated to field stays, and due to the sheer size of the park. Although support exists for a participatory approach, reluctance remains to rely more fully on the community perspectives and knowledge identified via participatory processes. Additionally, the high levels of external funding are scheduled to end in 2007. This is of concern as the participatory methods are still being implemented. The park team has found that many methods (for example, the MUF) have been expensive to implement, both monetarily and in terms of time allocation. Finally, pressures on the region's resources are tremendously high. Immigration threatens to swamp efforts at local level sustainable resource use. The heterogeneous nature of immigrant communities on the park's western edge makes participatory approaches to conservation even more challenging. Illegal resource use activities (logging, land sales, poaching) are being carried out by some local families and by external agents. Many illegal resource users are not participating in the management process, and some are even lobbying against collaborative management efforts.

## Challenges and opportunities for using Participatory Action Research for collaborative resource management

The case of the Cordillera Azul National Park demonstrates the role that PAR can play in removing some (but not all) barriers to more collaborative and community-based resource management. The work carried out by communities and researchers during the ethnobiological study and the MUF demonstrated local resource needs, mapped resource use patterns, provided evidence of local conservation norms, built local community capacity to monitor and manage resources, and provided more secure land rights. The ultimate success of such a participatory conservation approach, however, still requires many other potential barriers to be overcome. The most imminent threats revolve around political stability (such as will INRENA cede control?), non-local resource users, and funding. These barriers are largely outside the sphere of influence of the locally-based PAR agenda we have outlined. Therefore, while PAR clearly has a vital role to play in advancing more participatory resource management, its limitations are equally evident. Ultimately, multiple factors at numerous scales influence the success or failure of collaborative or community-based conservation (Sunderlin *et al.* 2005), and PAR is best able to tackle those factors that are most local in origin. The successes in the Cordillera Azul have taken place in a range of different communities, varying in ethnic composition, livelihood strategies, ecological constraints and level of connectivity with other communities, demonstrating the flexibility and effectiveness of participatory methods in a diverse array of local scenarios. A PAR approach may be a means of overcoming some programmatic failures that have plagued community-focused conservation strategies; however, long-term viability will still require adequate funding and sustained support from stakeholders at all levels.

## Notes

1 Until 2002, this was APECO; in that year, a small group of APECO staff created an independent NGO – *CIMA-Cordillera Azul* – which continued with the park conservation efforts.

2 I.e. Field Museum, including Alaka, and *CIMA-Cordillera Azul* staff, including Miguel.

# 9 Researching sexual health

## Two Participatory Action Research projects in Zimbabwe

*Mike Kesby and Fungisai Gwanzura-Ottemoller*

### Why take a participatory approach to researching sexual health?

A Participatory Action Research (PAR) approach can be extremely useful to those working in the complex and difficult area of sexual health, and this field illuminates the wider practical, ethical and political advantages and challenges of PAR (see Gordon and Cornwall 2005). Therefore, before discussing specific techniques, processes and challenges, we review the specific utility of PAR.

Poor communication is a major element of Africa's HIV crisis, and many people are reluctant to talk about sex publicly or privately: 'Sex is happening but people are not talking about it' (workshop participant cited in Kesby 2000a: 1723).

Researchers often therefore find this topic and relevant social groups 'hard to reach'. However, difficulties frequently arise as much from researchers' methodologies as from anything inherent to issues or respondents. The study of others' sexuality is certainly a minefield whether one is an 'insider' (for example, Zimbabwean like Fungi) or 'outsider' (British like Mike), and requires considerable prior knowledge of a context if judgmental interpretations are to be avoided. Our experience is that people often want to 'talk about sex' but lack the opportunity to do so, either in everyday life or in conventional research (Kesby 2000a; Gwanzura-Ottemoller and Kesby 2005). In our work on sexual health and HIV in Zimbabwe, we have found that PAR's avoidance of formality and hierarchical relationships, the involvement of participants in research design, the emphasis on hands-on methods, self-analysis and recognisable action outputs has enabled us to work across axes of difference and to engage adults, young people and children in deep discussion about very personal issues. The techniques we have used help to get difficult issues 'out in the open' (Kesby 2000b), while the ground rules and other participatory mechanisms that frame our research help to create socio-spatial arenas in which participants can talk about sex in relative safety (Kesby 2005; see also Kesby *et al.*, Chapter 3; Adams and Moore, Chapter 6; and Alexander *et al.*, Chapter 14 in this volume).

A second reason for employing a participatory approach in sexual health research relates to ethics (see also Manzo and Brightbill, Chapter 5 in this volume). When research identifies beliefs, behaviours and knowledge deficits that directly threaten the lives of respondents, should our aim be to leave the research context undisturbed? Social research is certainly filled with moral hazards for investigators and actual risks for participants; assumptions based on the supposed normality, superiority and homogeneity of a 'European model of (hetero)sexuality' must be avoided (Kesby *et al.* 2003). However, in the spirit of activism (see also Chatterton *et al.*, Chapter 25 in this volume) it should be possible to share our considerable knowledge about sexual health and our skills of social investigation without imposing 'norms' (Milofsky 2006: 467). Rather, PAR opens a space in which collaborative analysis can emerge between researchers and researched in the field in real time, not only the virtual time–space of text. Thus we feel strongly that in a context where some 1.7 million are HIV positive in Zimbabwe, prevalence rates among 15- to 49-year-olds exceed 20 per cent, and average life expectancy is about 39 (WHO 2006), the ethical maxim 'do no harm' is insufficient. A PAR approach can help shift the ethical basis of research toward 'doing some good'.

Thirdly, PAR's underlying ontology, the belief that human beings are dynamic agents capable of reflexivity and self-change, is well suited to the study of health related behaviour. Conventional health research seeks to facilitate self-change but separates this activity in time and space from the process of data collection and analysis. Moreover, it tends to conflate knowledge acquisition with behaviour change and so feeds into policies that deliver 'correct knowledge' as a technical package. By comparison, rather than making human agency simply the object or target of experts' research, PAR provides an epistemological approach that accommodates the reflexive capacities of human beings within the research methodology itself. Participants explore their experiences and beliefs, analyse data generated through research and come to their own conclusions about necessary action in their communities (Pain 2004; see also Kindon *et al.*, Chapters 1 and 2; Kesby *et al.*, Chapter 3; and Pain *et al.*, Chapter 4 in this volume). However, agency (especially sexual decision making) must not be seen as individualised and autonomous but rather as socially embedded, contingent and achieved using available resources. Such an ontology makes a participatory epistemology even more relevant to sexual health: PAR methodologies which encourage group analysis of the relational nature of socio-sexual phenomena can provide rich data on the factors that put people at risk, and provide participants with resources to reformulate their agency in more empowered guises (Kesby 2005; Kesby *et al.*, Chapter 3 in this volume). Finally, PAR can help catalyse, immediately and locally, the kind of changes that other health research sometimes claims to attempt as a long term goal.

## Some participatory techniques for sexual health research

In this section, we review some of the techniques used in our own research.

### *Participatory 'KAP' questionnaires*

Questionnaire surveys exploring 'knowledge, attitudes and practice' (KAP) have often been the tool of choice in sexual health/HIV/AIDS research and have been used to map broad patterns across continents. However, they seem inherently non-participatory: often predetermining the range of answers available, involving little participant input in design, and mining localities of raw data which are taken to be processed elsewhere. Positive impacts on respondents may not be entirely lacking, but given the resources involved, surveys often represent a missed opportunity for learning in researched communities themselves (see Stuttaford and Coe, Chapter 22 in this volume).

Nevertheless, when conventional techniques are deployed through participatory frameworks, their politics and impacts can fundamentally change (see, for example, Opondo *et al.*, Chapter 10; Hume-Cook *et al.*, Chapter 19; and Elwood *et al.*, Chapter 20 in this volume). Firstly, questionnaires can be jointly designed with collaborators and participants (Gwanzura-Ottemoller 2005; see also Macfarlane and Hanson, Chapter 11 in this volume), so that questions address issues of local concern and use local terminology to which participants will be more responsive. Secondly, questionnaires can be deployed rapidly, in groups and in the absence of literacy. For example, questions can be read aloud, with participants responding privately on a 'ballot style' questionnaire that features only a list of numbers and labelled columns (for example, '✓', 'X' and '?' representing 'yes', 'no' and 'don't know'). Thirdly, the advantages and value of quantitative data (such as anonymity, 'hard numbers', descriptive statistics, etc.) can be made available to participants in the field. For example, researchers retrieving simultaneously completed questionnaires can undertake rapid provisional analysis of key questions such as 'Can you get pregnant the first time you have sex?' and engage participants in immediate discussion of dangerously incorrect knowledge. More generally, the 'power of numbers' can be made available to participants for self-analysis and for action in (and beyond) their surveyed community.

### *Diagramming as a means to get issues out in the open*

Alternative visual methods are tremendously useful. By physically getting difficult issues 'out in the open', they make them easier to discuss (see also Opondo *et al.*, Chapter 10; and Alexander *et al.*, Chapter 14 in this volume). Below we discuss two examples.

#### *Diagramming 1: flow diagrams*

In our research, various forms of flow diagram have proven a powerful means to research sexual health and facilitate participants' self-analysis. When Mike worked with groups of unmarried Zimbabwean young people, he knew their understanding of HIV/AIDS was fairly good. The key question became whether and how such knowledge could be acted upon in the real life situations in which life-saving sexual health decisions had to be made. Diagramming enabled 'brain-storming' of a whole series of issues around 'contexts for sex' (see Figure 9.1); participants added and rearranged cards, made connections between elements and scored scenarios according to their prevalence.

Once certain known but little discussed facts were 'out in the open' it became possible to 'interview the diagram' and engage in very frank discussion about sex (see Box 9.1).

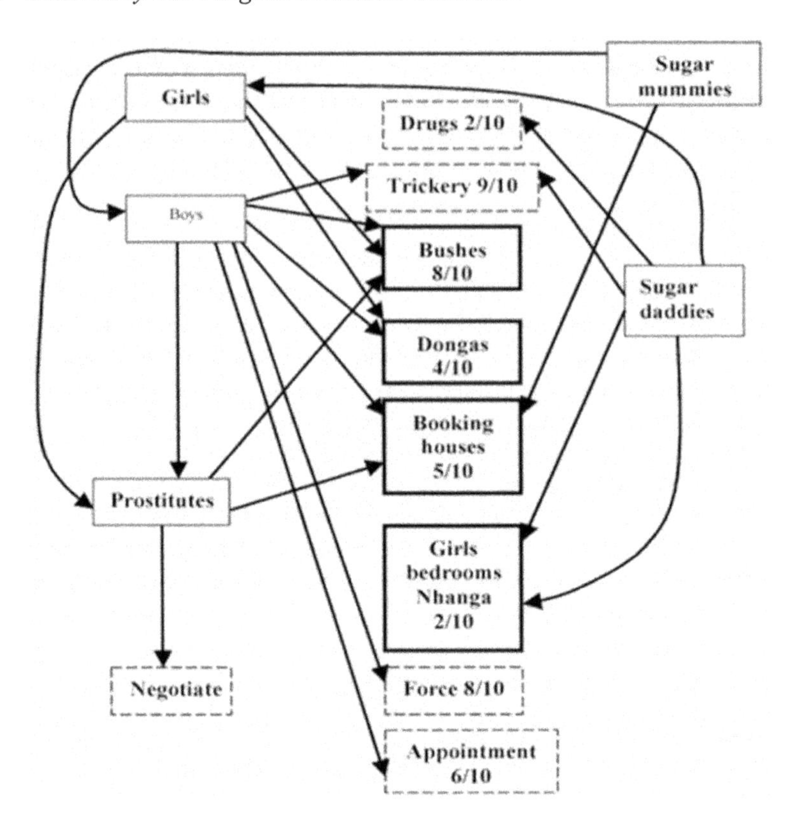


*Figure 9.1* Common contexts for sex (produced by young men who participated in an NGO HIV education programme)

## Box 9.1 Interviewing the diagram in participatory sexual health research

MIKE: Can you rate [these phenomena] according to your own experience; what you yourselves are doing.

YOUNG MAN 1: Trickery is mostly used ... because it is difficult to just negotiate so it needs a bit of trickery to get what you want

YOUNG MAN 2: The first time, you can trip them, but it is like an art form, you use a bit of force to get what you want. If you are hoping she will get carried away by nature but she doesn't and runs away, so you know that you have to catch her and trip her and take her or that will be the end of the affair.

Source: Group diagramming interview 2004

These young men had previously attended a participatory HIV education programme run by a local NGO that included sessions on gender awareness and rape. The young men were aware of the contradictions between this knowledge and their behaviour but lamented: 'but once you are out there, one-on-one, it is difficult' (Group diagramming interview, 2004).

Adults' reluctance to accept the sexuality of unmarried young people was forcing them to engage in sexual activity in marginal spaces like the dongas (dry river beds) and bush (wilderness areas). Consequently, sex was often hurried and condoms were difficult to negotiate, especially when 'a little bit of force' was seen as 'necessary'. Diagramming sessions showed that knowledge alone is not enough for safe sex decision making, highlighted the need for education programmes to address real life situations 'out there', and identified the difficulty of distanciating empowered performances beyond carefully controlled participatory arenas (Kesby 2005, 2007a; Kesby *et al.*, Chapter 3 in this volume).

### *Diagramming 2: body maps*

Body mapping is a lateral extension of the community mapping long used in PAR to enable participants to represent their social and physical landscapes. The human body has been extensively mapped by western science, but the ubiquity and utility of scientific physiology cannot simply be assumed; it is culturally specific (for example, it disassociates reproductive capacity from the divine), and may not be understood or accepted exclusively by people with 'traditional' beliefs, or westerners interested in homeopathy, faith-healing, etc. Cornwall (1992) pioneered body mapping as a means to reveal and valorise local anatomical conceptions and terminology and bridge scientific and vernacular idioms in ways that 'make sense' in local terms.

Fungi utilised body mapping in her investigation of sexual knowledge and behaviour among Zimbabwean school children (working in partnership with schools and a local NGO who could provide qualified counsellors if children reported abuse). Children are remarkably mature and reflective when using cartoon body maps, so long as groups and discussion are well handled. Body mapping was preceded by participatory questionnaires and other diagramming sessions, which had already opened discussion on sex. Fungi asked children to produce drawings that would help show 'where babies come from' and how 'HIV is transmitted'. Although their visualisations do not show much detail (see Plate 9.1), participants used them as a shared reference point through which to open up wider discussion on difficult topics.

The exercise raised many important issues: while younger children happily engaged in the task, they needed to be given euphemisms for the genitals because of embarrassment about using the terms 'penis' or 'vagina'. Moreover, some grade five (10 to 11-year-old) girls lacked confidence and understanding of anatomy using either vernacular or scientific lexicons, even though they had spoken confidently about HIV/AIDS while diagramming. Meanwhile, grade seven children (13 and 14-year-olds who learned human biology in sixth grade) happily used English anatomical words when body mapping but were embarrassed using Shona terms

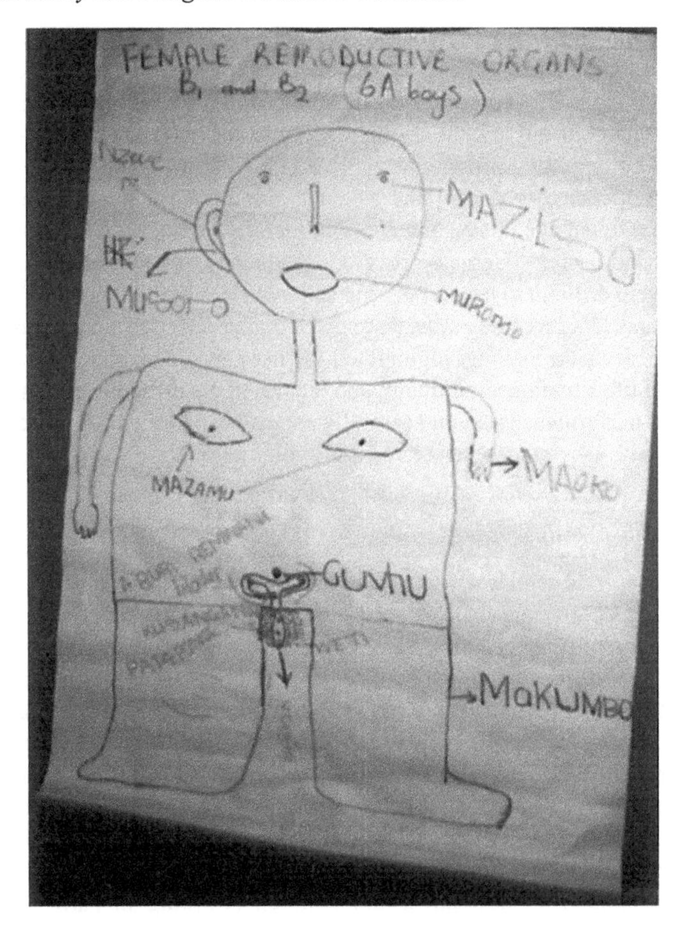


*Plate 9.1* Body map (grade six boys) (Credit: Fungisai Gwanzura-Ottemoller 2003)

loaded with sexual meanings. Finally, scientific language was used in conjunction with customary concepts: for example, men make babies, women are merely vessels that nurture embryos (Kesby 1999b). 'When the man produces sperms … they go into the woman, and … group together. Where they are gathered; that's where they'll produce an egg' (sixth grade boy quoted in Gwanzura-Ottemoller 2005: 200).

Body mapping enabled Fungi to explore the different registers and resources that children use and draw upon in their attempts to make sense of sexuality. It revealed that the current didactic curriculum on sexual health in Zimbabwe may not enable children to join up the various knowledge repertoires they possess. It structures HIV/AIDS and anatomy as a biomedical or academic realm of knowledge made in and through the classroom, leaving sex and the body as a realm of practice and pleasure made in and through the home, peer gossip and (sooner or later) lovemaking.

Body mapping (and PAR more generally) provides resources and creates spaces in which participants' agency can be reworked, connecting various knowledge repertoires in ways that enable them to better protect themselves (and others) from HIV infection. Certainly, this work is political and potentially 'normative'. For example, Fungi's decision to finish sessions by discussing 'anatomically correct' drawings favoured the scientific model and undermined the validity of vernacular knowledge. However, traditional notions of biological reproduction effect a powerful patriarchal appropriation of female reproductive capacities (Kesby 1999a) and discourage condom use. Thus, despite the power relations enmeshed in diagramming (see Alexander *et al.*, Chapter 14 in this volume), it nevertheless provides participants themselves with an opportunity to analyse the potentially dangerous absences and contradictions in their own understandings.

### *Drama: linking high theory with grass roots action*

Drama and role play are increasingly used in research (Gordon and Cornwall 2005; Opondo *et al.*, Chapter 10; and Cieri and McCauley, Chapter 17 in this volume). Experimenting with drama as an addendum to diagramming, Mike discovered that of all alternative methods he used, young people responded most enthusiastically to this one (a social genre already well established in this context). As a method of data collection, improvised dramas enabled youths to explore the real life contexts and decisions that expose them to HIV risk and sexual exploitation. As a method of feedback and dissemination, plays enabled youths to grandstand their knowledge of HIV to the whole community (exposing myths around the care of those living with AIDS and demonstrating best practice). Active participation in the feedback session (at which Mike also presented provisional results to the community) gave young people confidence to make a public appeal to the older generation to condone their use of condoms (see Van Blerk and Ansell, 2007 for an excellent discussion of participatory feedback and dissemination).

In future work Mike intends to explore the potential overlaps between this accessible form of participatory praxis and less accessible yet important theoretical insights. When working with peer groups it was striking how 'realistically' girls performed as boys (and vice versa) (Plate 9.2). It seems possible that '~~childish~~ methods'[1] like play-acting might offer a straightforward practical means by which ordinary people could perceive the 'made-up' or performative nature of all social identity and behaviour (see Kesby 2007b). This might help open discussion about the possibility of acting differently when negotiating sexual relations. For this to work, improvisations would need to do more than generate cautionary tales of promiscuity and its consequences (of the sort thus far recorded by Mike). Participants could be encouraged to explore the words, gestures and performances that might, in real life contexts, empower them to avoid and negotiate risk. In this way, participatory data collection might become a transformative rehearsal for reality, while participatory dissemination could provide audiences with useful scripts to recite in their own lives. Both activities would help provide people with resources that could help sustain empowerment beyond the carefully managed arenas of participatory projects (see Kesby *et al.*, Chapter 3 in this volume).

*Plate 9.2* Gendered performance: the youth on the left 'acts like a girl' (Credit: Mike Kesby 2004)

## Challenges

The particularities of sexual health research give rise to many challenges, and these are of relevance to PAR more generally. Firstly, notwithstanding the HIV/AIDS crisis, researchers must avoid simply imposing an agenda. People often have many more immediate priorities (for example, basic survival, but also desire and love) and unless these are recognised and/or addressed in parallel, people will not take ownership of a sexual health agenda.

Secondly, while power is obviously at work in neo-conservative programmes prioritising 'abstinence', fidelity in marriage and condom use only among 'high risk' groups (see US Presidential Emergency Plan for AIDS Relief; Gordon and Cornwall 2005), all sexual health work must be recognised as entangled with different dynamics of power. Liberal PAR initiatives can also have normalising effects: researchers often medicalise sexuality, and yet workshops exploring 'what makes a good lover' might be a better way to develop resources useful to the constitution of an agency capable of negotiating safer sex. Meanwhile, group discussions can generate stereotypical generalisations, fail to record diverse experience and practices and exclude taboo sexualities not discussable in such 'public' forums (Gordon and Cornwall 2005; Cornwall 2006b). Yet exploration of 'marginal sexualities' can be key to unlocking broader social change (see Phillips 2004). This said, whether our focus is 'safe sex' or 'pleasure', PAR interventions will be entangled with power and we should therefore anticipate some resistance (Kesby *et al.* 2003). Engaging with this resistance productively, rather than being frustrated by it, will ultimately help strengthen PAR projects.

Thirdly, researchers should avoid the inappropriate imposition of scale on sexual health projects (Mohan and Stokke 2000). While participants can do much to tackle HIV/AIDS locally, this disease is not a community scale phenomenon. Rather than repeating the mistakes of conventional health research in hoping that self-knowledge alone will initiate behaviour change, advocates of PAR can recognise the wider structural causes of the pandemic and seek to facilitate participants in making analytical and practical connections between local experience and national and global issues and action.

Finally, researchers need to think hard about the spatial dimensions of PAR. When the topic under investigation is sexual health, it is particularly evident that facilitating open discussion and empowerment within a carefully managed research project is only the first vital phase. With participants, researchers need to identify resources that can be distanciated beyond the research arena, so that they can be used to effect empowered forms of agency and decision making within the everyday contexts where sexual activity actually takes place (Kesby 2005; 2007a).

Despite these caveats, we believe that where PAR approaches are used sensitively and appropriately they not only provide rich data and valuable insights, but also help participants to develop resources valuable to their ongoing struggles for sexual health.

## Notes

1 Kesby (2007b) uses a strikethrough to signal that the normally derogatory meaning of the word 'childish' needs to be challenged and reworked.

# 10 Gender and employment

## Participatory social auditing in Kenya

*Maggie Opondo, Catherine Dolan, Senorina Wendoh and James Ndwiga Kathuri*

### Introduction

Satisfying the ever-demanding tastes of global consumers has led supermarkets and department stores to source products from farms and factories scattered across the globe. Today global supply chains carrying fresh products (such as cut flowers and fresh vegetables) and processed ones (such as garments and tea) are a defining characteristic of production spaces, particularly in countries in the majority world where such chains provide important opportunities and where certain sectors (eg tea in kenya) have benefited from integration in global supply chains that link European consumers with majority world workers and farmers. These sectors are not only the leading foreign exchange earners but also provide substantial employment opportunities for the millions of people.

Although participation in global supply chains provides new opportunities for countries in the majority world, they are also sites of acute risk and vulnerability. Actors in these chains face an increasingly precarious business environment as trade liberalisation creates fierce competition and ever-declining prices. To reduce costs and improve market position, producers and buyers seek ever-greater flexibilities. As a result, flexible labour strategies – 'non standard' work arrangements such as casual, seasonal and/or temporary and contract labour (Storper and Scott 1990) – are now a defining feature of global supply chains. However, the advantages of flexibility to one agent in the chain often come at the expense of others. Exporters can accommodate uncertainties in supply and demand by shifting the risks of production onto a flexible workforce that can be rapidly drawn in and out of production, thereby avoiding many of the non-wage costs of employment (Standing 1989). However for workers the outcome is less sanguine; they lack security of employment, have few employment rights, receive inadequate employment benefits or social protection and lack trade union organisation (Barrientos *et al.* 2003). In response to increasing public concern, growing numbers of retailers, manufacturers and importers are applying codes of conduct to ensure minimum labour standards throughout their supply chains. These codes have become especially widespread in certain consumer goods sectors, such as fresh produce and garments, and are now widely applied in Kenya (Diller 1999).

In this chapter we examine how social science research methods and participatory techniques can be used to implement codes of conduct through a participatory framework known as Participatory Social Auditing (PSA). We begin by providing an analysis of gendered employment in the garment, cut flower, tea and fresh vegetable supply chains in Kenya and examine the extent to which the participatory approach used to audit codes of conduct has been successful in capturing workplace-related concerns. Drawing on case studies from these sectors, we argue that the use of a participatory approach is particularly useful in identifying the conditions of vulnerable and marginalised workers and in providing a vehicle for worker empowerment.

## Gendered employment

Over the last two decades the rising demand for low cost flexible labour has led to the employment of millions of women, particularly in the production end of global supply chains. In the majority world, the search for greater labour flexibilities has sparked a fierce race to the bottom as countries compete to attract foreign investment by promoting themselves as low wage havens. Yet low cost, flexible labour is not only spatially concentrated but it is also gendered. It is women who comprise the majority of workers in factories that pursue the low-road strategies of the new economy, as they are perceived as better suited for quick labour intensive work and less costly in terms of wages and non-wage benefits (Collins 2003).

As recent literature on female labour in global assembly lines highlights, the 'ideal' worker is not simply found but actively created through strategies of gendering that constitute the 'dexterous', 'compliant' woman as the preferred instrument of capital accumulation. Production is structured through gendered assumptions that create a pronounced demarcation between men's and women's work, with women often concentrated in the most labour-intensive segments of the production process (Dolan *et al.* 2003). For example, women in the fresh produce sector are engaged in picking, packing and valued added processing, while men are primarily engaged in spraying, irrigation, construction and maintenance. Gender segregation is thus characterised by asymmetrical power relations, with men occupying the senior and better-remunerated permanent positions while women are concentrated in the more insecure, temporary, and typically lower waged positions.

The pursuit of cheap and flexible labour is often touted as being in the best interest of the most vulnerable – the unskilled and unemployed – who face few alternative income earning opportunities. Yet women workers in global supply chains face difficult working conditions due to their predominance in the most labour intensive aspects of production. Key challenges include: insecure forms of employment; compulsory and excessive overtime work; sexual harassment; low wages; lack of collective bargaining power; exposure to occupational health hazards and restriction of reproductive rights. During the 1990s, these conditions galvanised consumer concerns when allegations of worker exploitation were circulated through the Internet and featured in the European press. With the prospect of regulatory sanctions and profit loss, European retailers began to reposition themselves as responsible corporations through a proliferation of codes of conduct, social standards, labels and guidelines.

## Codes of conduct

With ethical issues moving to the forefront of corporate concerns, a burgeoning number of voluntary codes of labour practice emerged in Kenya to address the production processes of companies exporting to minority world markets. These codes, which aim to ensure minimum labour standards by setting guidelines on a range of workplace-related issues, are largely based on International Labour Organisation core conventions and the United Nations Declaration of Human Rights (Dolan *et al.* 2003).

How codes are implemented, monitored and verified directly determines the extent to which they will improve labour conditions for workers in global supply chains. One way to overcome the pitfalls associated with top-down code development is to ensure that workers' voices and concerns are included in the auditing process. Standard auditing practice focuses on assessing records and responses of firms and managers on the working conditions of mainly permanent workers. Worker interviews are usually brief and not sufficiently inclusive of all categories of workers. Such a snapshot approach to auditing proved insufficient in capturing sensitive workplace issues such as gender discrimination, sexual harassment and abuse, and provided us with only a partial (and often biased) rendering of workplace concerns (cf Auret and Barrientos 2004; Dolan *et al.* 2003). These inadequacies prompted our search for a new approach to identify on-the-ground labour practices and capture the experiences of men and women working in global supply chains.

## Using a participatory approach in workplace research

Participatory Social Auditing (PSA) adapts and applies standard social science methods (for example, semi-structured interviews (SSIs) and focus group discussions (FGDs)) and participatory techniques (such as transect walks, group mapping and ranking and role play) into the auditing process through a participatory framework. It engages more directly with workers and worker organisations in the process of code implementation and assessment. PSA is centred on a process approach which seeks to instil learning and improvement through management and worker education rather than simply checking for one-off compliance (see also Stuttaford and Coe, Chapter 22 in this volume). By fostering partnerships between different actors (for example, companies, trade unions, NGOs), PSA aims at developing a locally sustainable approach to the improvement of working conditions (Auret and Barrientos 2004).

Between 2002 and 2006 we undertook three separate studies (flowers, fresh produce and tea and garment sectors) that used a participatory approach to assess how best to identify gender-sensitive information and encourage workers to participate in workplace improvement within global supply chains. These studies showed that such an approach can create new spaces for addressing the employment concerns of marginalised workers whose voices are typically overlooked in conventional audits. For example, the discriminatory practices embedded in the recruitment process of the garment sector were highlighted in a role play in which workers imitated Chinese managers during recruitment interviews saying, 'Here big, there big, no good' (pointing to the breasts and behinds of the female worker interviewees). Hence, during the first stage of job interviews the interviewees were requested to parade themselves and if one was perceived to be 'big here, big there' then they were automatically disqualified from obtaining employment (KEWWO 2006).

In our experience, a participatory approach more easily allowed for the exploration of sensitive gender issues that emerge in feminised supply chains. This was apparent in our FGDs, in which a number of sensitive issues emerged – sexual harassment, abortion and sexual favours – that would have been unlikely to surface using more structured methods, which do not allow for the voice of the marginalised to be heard (see also Kesby and Gwanzura-Ottemoller, Chapter 9 in this volume). In a FGD on a flower farm, workers revealed that abortions are common because pregnant women in temporary positions are typically not given another contract when their current one ends and are the first to be dismissed in case of a *mchucho* (when there is no work in the company). A group of casual female packhouse workers lamented, 'Many female workers have had abortions, even at eight months to avoid the risk of being sacked' (Dolan *et al.* 2003: 49).

By involving the poorest and most powerless, a participatory approach can also initiate a process of empowerment among workers. The interaction and group dynamics of participants in the FGDs, for instance, not only facilitated group cohesion but also provided a safe space for them to freely express their concerns. The defiance of the female garment worker who would not exchange sexual favours for bonuses is an indication of how group exercises encouraged workers to verbalise their grievances and enhance their confidence (see also Box 10.1).

In addition, we found that the camaraderie that evolved among different groups of workers engaged in a FGD was a potential source of empowerment. For

**Box 10.1 Role play: use of abusive language and inhumane treatment (garment sector)**

The skit begins when everyone is at their work station and the supervisor is walking round. He discovers that a worker has put her head down on the machine; he then taps her on the head and asks, 'What is wrong with you? It is 12.00 noon, you are about to go for lunch and you have not even done a quarter of your target'. She says that she is not feeling well. She then asks the supervisor for permission for the third time to go and see a doctor, the supervisor declines and abuses her that she goes out every weekend, sleeping around with several men, so her mind is only fixed on sexual performance and not the work. 'So you come here to make bad clothes. If you do not want to work you may leave now, there are a number of job seekers at the gate who would like to fill this position'. The supervisor then tells her to go home for good if she is sick.

Source: KEWWO 2006: 26

example, in one FGD a worker in the flower industry who had rejected the sexual advances of her supervisor was able to share her fears of being dismissed and to draw compassion from her fellow group participants (Dolan *et al.* 2003). FGDs were also used to create worker awareness of the rights and protections that should be afforded them in global supply chains (see also Chatterton *et al.*, Chapter 25 in this volume). In all four sectors – cut flowers, fresh vegetables, garments and tea – our SSIs revealed that most workers had little knowledge of codes or the nature of the supply chains in which they worked. Thus during the FGDs, facilitators provided details on the codes and also engaged the workers in group exercises to map out supply chains.

A participatory approach was flexible enough to accommodate differences in workspaces and to engage with both formal and informal workers. For example, while codes often extend to a company's first-tier suppliers and their permanent employees, they rarely cover all workers in the supply chain and generally exclude small producers who are prevalent in the fresh produce and tea sectors. Such conditions often mean researchers are unable to elicit information from all workers and producers in global supply chains. Participatory techniques can ensure that the concerns of less visible workers, the majority of whom are female, are heard.

## Lessons learned from field experiences

A participatory approach is a powerful instrument for workplace research that aims to achieve a deeper understanding of gendered labour conditions. It facilitates the engagement with different types of workers regardless of their position in the supply chain, and the collection of data from formal and informal workspaces. However, adapting standard social science methods within a participatory framework is not without its challenges. For example, SSIs are not well equipped to deliver a more comprehensive portrait of working conditions while female workers tend to be more apprehensive and less confident in their responses than their male counterparts. We found that to garner more than a 'yes/no' response from women we had to continually prompt and encourage them to elaborate. Yet since there is often little opportunity to probe in depth, there is a high risk that the responses will only provide a partial understanding of workplace issues. Since SSIs are time-consuming they can interfere with workers' ability to meet their obligations at home or at the workplace, workers may attempt to hasten the interview process by providing curt and/or compliant responses, which may not always reflect the reality of their situation. Finally, the individual interview can be an intimidating experience for workers who, despite the assurance of confidentiality, are likely to worry that disclosing certain information could cost them their jobs. It is therefore important to combine SSIs with the use of other participatory techniques (Dolan *et al.* 2003).

The information from SSIs was a useful springboard for a more in-depth exploration of workplace issues in FGDs. In the FGDs we found a number of pertinent issues emerged such as workers' perceptions towards their working environment; their relationships with the supervisors and management; sensitive matters of sexual harassment and sexual favours; and the degree of empowerment experienced among the different groups of workers. However, the effectiveness of FGDs was often highly constrained by the nature of power relations within the group. For instance, certain group members tended to dominate discussions and inhibit the less articulate members from contributing. This was particularly the case in mixed-sex FGDs, where there was a tendency for women to defer to the more dominant male participants. Our experience shows that in a research setting, it is especially important that the researcher carefully considers the composition of the group, and its likely impact on the substance of the discussion (Dolan *et al.* 2003).

Our research projects also explored the effectiveness of participatory mapping and ranking exercises, role plays, and transect walks for ascertaining workplace information. During the mapping, workers began to develop a shared vision of their workplace, which emerged through lively discussions of where particular people and production activities were situated. This technique was also effective at capturing gender differences in the perception of workspaces. For us, mapping provided entry points for developing rapport with workers and the production environment. However, mapping was a time-intensive exercise and required that both researchers and participants were committed to the process (Dolan *et al.* 2003). Similarly, transect walks provided insights into both social relations and working conditions (Auret and Barrientos 2004). For example, it was only through the transect walk of a flower farm that we were able to experience the intensity of heat and humidity of the greenhouses.

Role plays were a particularly valuable tool for acquiring a deeper understanding of certain workplace issues (for example, harassment, abuse, discriminatory hiring practices), for promoting group cohesion, and for eliciting greater participation of reticent group members (see also Kesby and Gwanzura-Ottemoller, Chapter 9, and Cieri and McCauley, Chapter 17 in this volume). Role plays also provided an opportunity for workers to reflect on and enact aspects of their employment situation which they may otherwise have felt reluctant to discuss. Women particularly embraced this activity and revealed more thoughts and attitudes (Dolan *et al.* 2003).

We found ranking techniques (Plate 10.1) to be more complicated for participants. Some participants were able to rank issues only after concrete examples were provided. Moreover, while they found it relatively straightforward to rank the first three issues, they thereafter found it difficult to determine the relative importance of the remaining issues (Dolan *et al.* 2003). Nevertheless, ranking techniques provided one of the few vehicles through which workers were able to prioritise workplace concerns, and offer researchers a lens into the relative importance of certain issues (Auret and Barrientos 2004).

We also faced certain challenges in deepening worker participation in the research projects because the context within which the research was carried out only allowed for a certain amount of indirect participation. Although we did not directly engage workers in all the stages of the planning design, their responses from the SSIs did determine the design of the FGDs. This notwithstanding, our research has contributed (in conjunction with efforts from the civil society

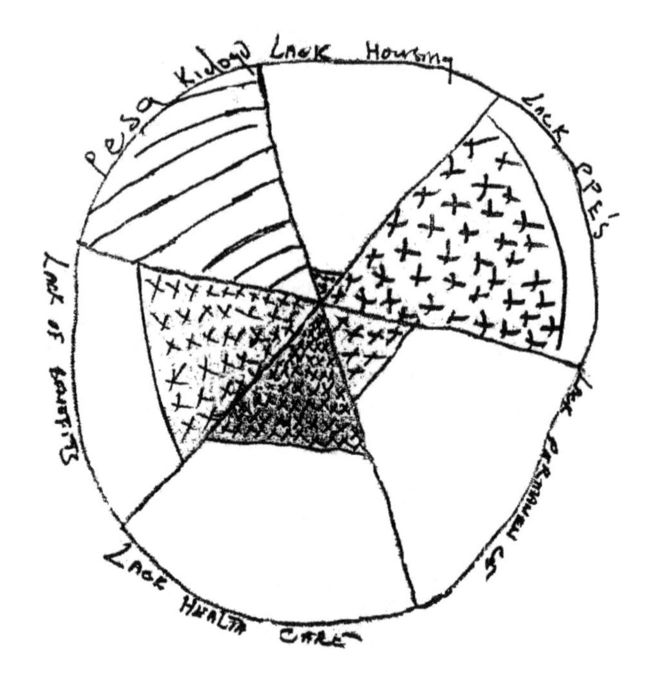


*Plate 10.1* Ranking the problems by tea male workers (Credit: Ethical Trading Initiative, ETI smallholder guidelines 2005) *pesa kodogo* = insufficient wages

organisations) to improved working conditions particularly in the cut flower and tea sectors. These include: increasing the number of permanent workers; awarding benefits to temporary workers; increasing attention to health and safety issues; reducing overtime hours; raising the housing allowance and providing workers with contracts. Overall, there is now a heightened awareness on farms of the need to improve the conditions of the workforce.

## Conclusion

Global supply chains draw large numbers of female workers in the majority world into labour-intensive production. While these chains offer positive opportunities for growth and employment they are also characterised by strong competitive pressures that force these industries to reduce costs through flexible labour strategies.

During the 1990s a plethora of codes of conduct emerged to counter growing consumer concerns over poor working conditions in global supply chains. However, the methods used to audit these codes often yielded little insight into workplace conditions. In contrast, a participatory approach offers an alternative to extractive information gathering and facilitates fuller understanding of how global supply chains differentiate the opportunities of men and women to engage in work, and shape their subsequent experience of that work.

In this chapter, we have described why and how a participatory approach (including standard social science methods and participatory techniques) can provide a clearer window into the lived realities of the workplace and prove useful to social auditing. Foremost, a participatory approach counters the hierarchical investigative approaches that often characterise the researcher–researched relationship, by eliciting workers' participation in and ownership of the research process. As our discussion shows, the systematic use of standard social science methods and participatory techniques *within a participatory framework*, can help to ensure the active involvement of different categories and genders of workers in the research process, particularly vulnerable and marginalised workers whose voices are often eclipsed by the standard deployment of survey instruments. By including workers in the process of research, this approach can enable workers to both define the circumstances that constrained them and support them to work out solutions for themselves. Since PSA is centred on a process approach, it enables learning, thereby creating a basis for more sustainable improvement (see also Stuttaford and Coe, Chapter 22 in this volume). In so doing, participatory research can help to ensure that the workers in supply chains such as tea, horticulture, and garments also partake in the much-touted benefits of global trade.

# 11 Inclusive methodologies

## Including disabled people in Participatory Action Research in Scotland and Canada

*Hazel McFarlane and Nancy E. Hansen*

### Introduction

This chapter discusses Participatory Action Research (PAR) undertaken while we were post-graduate students in the Department of Human Geography at the University of Glasgow, Scotland. We discuss the influence of our activism in the disabled people's movement in relation to our politicised view of disability as a socially constructed form of oppression. We also reflect on the way that this concept of disability played out in the implementation of inclusive research methodologies.

In the following section we situate ourselves as academics and activists, describing the potential of PAR to engage critically in the kind of emancipatory research approach advocated by, for example, Mike Oliver (1997). We then provide an overview of the research process, focusing upon design, data generation and action outcomes.

Our interests in PAR were several: PAR is often successful in making links with 'hard to reach', marginalised groups (Kesby *et al.* 2005), and thus seemed an appropriate way to work with disabled women on their experiences of motherhood and employment. PAR encourages the negotiation of power relations and gives added weight to the voice of the researched (Kitchin and Tate 2000). A participatory approach allowed us to extend our theorisation of disability, as a social construct, into methodological practice, by consciously seeking to accommodate our respondents' needs in our research design. Furthermore, PAR appealed to us as a means to describe and interpret women's social reality, providing potential opportunities for modest but meaningful change for disabled women.

### The academic-activist: potentialities of PAR

Prior to returning to academic study we were involved in disability activism in Scotland and Canada respectively. Our involvement in campaigns for disability civil rights legislation and demands for radical social change acted as a catalyst that prompted our return to university as academic-activists (Routledge 1996). We saw PAR as an opportunity to engage disabled people in the research process (identifying and addressing issues of importance in their lives; gathering, documenting and representing their embodied experiences), thereby challenging disabling barriers and effecting change through collective and/or individual action.

Involvement in disability activism afforded us an in-depth understanding of the Social Model of Disability (Oliver 1992). Consequently, we reject the 'traditional' or medical model of disability, which identifies an individual's bodily impairment as an abnormality requiring correction and normalisation. In contrast, the social model rejects individualised, deficit-orientated understandings of disability and locates the phenomenon within its social context. From this perspective it becomes clear that it is the socially produced environment, social barriers and oppression that disable people, not their impairments as such. Inaccessible buildings and public transport, demeaning stereotypes, prejudice and ignorance, and negative social ascriptions, undermine and deny the many capacities of disabled people. Human Geography research lends itself to the social model's approach in that it combines the model's focus on physical and attitudinal barriers with an appreciation of the construction, uses and meanings of public and private space.

Our personal experiences as disabled women informed our research. Hazel's decision to study disabled women and socio-spatial barriers to motherhood, a project involving twenty-seven women resident in Glasgow and Edinburgh (McFarlane 2005), was influenced by her experience of hostility in the reproductive arena. Similarly, Nancy's study of disabled women's access to employment arose from personal experience. It involved twenty women residing in Scotland and twenty in Ottawa, Canada (Hansen 2002).

Traditional research methods construct the researcher as 'expert' and treat the researched as *objects of* research, rather than subjects *with whom* research is conducted. Where the 'target group' is disabled, the former approach only serves to further marginalise people who are already objectified through other social processes. Mike Oliver (1997: 15) contends that research concerning disability issues that does not actively involve disabled people is nothing more than a 'rip off', unlikely to represent accurately issues of importance to disabled people, challenge their oppression or change the realities of their lives. In our research disabled women's involvement was crucial: an inclusive participatory approach offered a means to encourage involvement by a group potentially suspicious or weary of research (Kitchin 2000).

As post-graduate students, our experience of PAR was marked by conflict: while our activist background drew us to include participatory elements in our research, we felt constrained by the need to conform to traditional PhD training methods (see also Cancian 1993). Consequently, participation was limited to research design, elements of data collection, reciprocity within the research process, and action outputs on an individual basis once the research was concluded. However, doing research differently has to begin somewhere; an inability to attain an ideal degree of participation need not prevent researchers from integrating participatory elements into their work (Kesby *et al.* 2005). Our partial participatory approach had a positive influence on the extent to which women shared their experiences, ultimately enriching the quality of the research.

Disabled people are a growing presence on university campuses and are, increasingly, the instigators, and not simply the subjects, of research. These changes have impacted upon disability research (of course, not all disabled academics study disability) and issues such as employment, reproduction, sexuality and motherhood have risen to prominence thanks, principally, to disabled women (Chouinard and Grant 1997; French 1994; Morris 1994; 1996; Thomas 1999, Wates and Jade 1999).

As disabled women activists and as geographers, we are acutely aware that our impairments, identified by the use of a long cane and crutches, respectively, segregate us as 'outsiders' from non-disabled society while simultaneously positioning us as 'insiders' in the disability community: we are part of what we research. While Ruth Butler (2001: 272) cautions against such self-identification, warning that readers may ascribe negative attributes to authors who describe themselves as 'disabled' or 'impaired', we do so to reclaim 'disability', that is, rework its meaning and challenge negative stereotypes. During our fieldwork we found that describing ourselves as disabled gave our participants an indication of our politics, thus making it clear that, like Kobayashi: '[We] do not use other people's struggles as the basis for [our] research; [we] use [our] research as a basis for struggles of which [we are] a part' (1994:78).

We made it clear that our self-identification was based on the social model of disability, not on normative understandings or corporeal typologies of impairment, and we often debated these different representations with our respondents.

Our background in activism allowed us to draw on a network of individual and organisational connections in Edinburgh, Glasgow and Ottawa: this enabled us to gain the support and endorsement of the five disability-led organisations[1] most prominent at the time, and permitted us to recruit more easily the users of services.

## Getting started

We wanted to ensure that the research would address, respectively, parenting and employment issues that were relevant and useful to the disability constituency and would address information gaps identified by disabled women. Thus we began by contacting members of frontline disability-led groups in Scotland and Canada with a view to soliciting their participation in our research designs.

We realised that if our commitment to developing an inclusive approach was to work in any meaningful way we had to anticipate potential barriers to women's participation. Thus, our second step was to form a research advisory group consisting of six disabled women. Together, we identified that significant obstacles to disabled women would be inaccessible information, transport and venues; inflexible personal assistance, and lack of affordable childcare, and we worked to find potential solutions to facilitate their participation. The advisory group suggested that if we showed potential respondents that we had identified and addressed such barriers, we would convey a clear message that their access requirements would be met, and that their participation, opinions and experiences would be encouraged, respected and valued. The research design process was a truly collaborative and mutually informative and transformative experience. For example, Nancy in particular had previously never 'allowed her disability to impinge on her academic work'. Thus it was a watershed moment in her own self-perception to be encouraged by participants to consider her needs in the interview context.

Funding arrangements and our student status determined that the subject area be established by us; however, the advisory group collaborated by discussing and identifying the parameters of the research. They agreed that research questions should explore disabling social barriers and not inquire about impairments or medical conditions. During discussions about methods the women concurred that, due to the sensitive nature of the research, focus groups would not be appropriate forums to discuss their experiences of access to reproductive choices and employment. They feared that, within close-knit disability communities like Edinburgh, Glasgow and Ottawa, participants would be known to each other and anonymity and confidentiality could not be assured (see also Mohammad 2001). Hence, the advisory group favoured one-to-one semi-structured interviews as an appropriate method.

## In the thick of it: doing Participatory Research

Participants were contacted via the five partner organisations. In an informal cover letter (in accessible formats: Braille, large print, tape, disc and electronic) we introduced ourselves as life-long disabled women and briefly outlined our research interests and commitment to a participatory approach, emphasising our willingness to facilitate participation on their terms. We also divulged our personal contact details and included a short article about our experiences. We shared this information to obviate any potential or perceived power hierarchies (Oakley 1981). We wanted to demonstrate that we shared many of our participants' struggles and did not perceive ourselves as superior (see Avis 2002). Many women identified with our short article and the sense of shared experience this engendered made it easier for them to recount their experiences.

Our endeavours to unsettle power hierarchies continued within the interview process: for example, we provided the women with a list of discussion topics prior to their interview. This enabled them to make choices about their degree of contribution and meant that respondents were in a more informed and, perhaps, powerful position than in conventional research encounters. We invited the women to determine interview times and locations. This came as a surprise to many, accustomed as they were to living in a society that is often physically and attitudinally hostile, and all appreciated greatly that the nature of their participation would be determined by them. We further shared control of the research interaction by agreeing, when invited, to stay for lunch. Formal interviewing moved smoothly into informal socialising, hosted and directed by the participants, their substantial commitment to answering our questions reciprocated by our respectful, trusting attitude. Thus, while the constraints of our topics and our post-graduate status obliged us to collect data from individuals using relatively conventional tools, we worked to empower participants in the research interaction.

Interviews were filled with mixed emotions: laughter was intermingled with sadness, regret, anger and frustration. The topics of employment access and reproductive choice meant that interviews were sometimes upsetting but they also proved cathartic. We experienced none of the research weariness noted by Kitchin (2000), with many respondents welcoming interviews as a rare opportunity to talk about their experiences. One woman remarked that participation had enabled her to end a thirty-five-year silence and that she actively wanted to share her experiences in order to 'make a difference for other disabled women' (McFarlane 2005). Clearly, our efforts to put participants at their ease, and ability to share similar experiences that let participants know they were not alone (Vernon 1997), paid dividends. Significantly, the result was not simply that 'deeper data' were gathered, but that participants felt actively engaged in a project to tell the story of disabled women's lives and effect change beyond their own circumstances.

## Action outcomes

Superficially, our two projects might seem to share more in common with deep qualitative research than PAR because our 'methods' were conventional and because research interactions and action impacts were predominantly on an individual rather than group basis. However, while we were somewhat constrained by our post-graduate status, we nevertheless felt that our projects captured key elements of a PAR approach. Not only did we facilitate a participatory research design phase and encourage shared control over the interview process, the projects also gave respondents the sense that while working individually they were participating in a collaborative project among and for disabled women. Furthermore, while often separated by time and space there were clear connections in our projects between participation in research and action outcomes.

Many of the women noted that sharing their experiences had helped to make sense of them. The opportunity to verbalise matters previously kept secret provided a sense of release, while recounting painful experiences helped lay them to rest. Furthermore, the effect of collaboration was to lessen the sense of social isolation felt by many of the women. In our interview discussions we conveyed (anonymously) the stories of others and each interviewee thus became aware that they were not alone in their experiences.

In and of itself such reflexivity through research was a positive action outcome for many of the women because taking control over one's emotional and experiential archive is a first step to a more empowered agency. This process was evident in the case of one participant who, through the research, reflected on and analysed her relationship, concluding that she was in an abusive marriage. The exchange of stories with the researcher and the discussion of the social model of disability were resources that enabled her assertion of agency (Kesby 2005). With a new perspective on the political in the personal she came to realise that her partner's interaction with her as a disabled woman was inappropriate, inequitable and unacceptable. Subsequently, she acted to assert her self-interest and left the abusive relationship. This example is an extreme instance of an 'action outcome' and researchers might normally view such impacts as 'unethical'. However, an 'ethics of participation' (see Manzo and Brightbill, Chapter 5 in this volume) allows that ordinary people, as competent individuals, will draw upon and learn from a research interaction as they see fit. While deep qualitative investigation of sensitive topics might galvanise reflection on problematic aspects of participant's lives, PAR goes further in providing solutions to problems.

In addition to reflection and action, facilitated *directly* through the research, our projects also generated other *indirect* outcomes. For research to be truly reciprocal, researchers must be prepared to get involved in matters 'not directly relevant' to the research (Vernon 1997). We followed up many interviews by either providing contact details for organisations or by supplying resources that we had discussed with the women. In some cases the interview itself represented a reciprocal arrangement because, at their request, we hosted some women for the weekend, interviewing them in our homes to give them a break from their domestic situations. Hazel provided one woman with assistance with an on-going housing issue, and the matter was resolved satisfactorily. Likewise, Nancy advised a group of women about opportunities in further education and two subsequently went to university. Thus, 'reciprocity in the research process' continued long after our 'projects' concluded.

While the constraints of our PhD programmes and the women's lack of interest in co-authoring academic publications has meant that much of our written output has been single-authored works, we nevertheless regard these as collaborative projects that reveal the lives and experiences of disabled women. It is our intention that they influence policy and practice in the areas of sexual health, family planning, maternity services and access to work schemes, thereby precipitating further action. Indeed, our projects have already contributed toward such effects in the UK: many of our participants were disability activists, involved in the campaign for improvements to the Disability Discrimination Act (DDA). While issues from that campaign influenced the design and implementation of our research projects, shared experiences from our projects also informed and strengthened the resolve of those involved. Since the completion of our research the Disability Equality Duty (DRC 2005, see www.drc-gb.org) has been implemented and some of our collaborators have been involved in ensuring that public bodies, such as the health service, promote the equality of disabled people. The activist and participatory dimensions of our approach made useful links between small-scale research and larger-scale movements and legislative processes, enabling participants to contribute to major social changes that will affect their and other's lives.

Two final, more directly 'collective action' outcomes of the research are worth noting. Notwithstanding their reluctance to commit themselves to the production of academic texts, our participants expressed an interest in forms of dissemination that may reach a broader audience. Thus Hazel has agreed to help two women write about their experiences of institutional life. A wider collective project is to follow up on a research question that asked if the women would be interested in developing a drama production based on their experiences. A number of women agreed to provide a further anthology of their experiences, which will be used to develop a stage production. The women will participate in the writing and development of the production alongside a professional playwright and actors. We see this as an exciting continuation and expansion of the participatory and activist elements of our, initially, more constrained academic research (see also Cieri and McCauley, Chapter 17 in this volume).

## Conclusion

Our research experiences as and with disabled women illustrate several important points. They reveal that tensions between conventional academic practice and dynamic activist commitment to meaningful change can be addressed by PAR (see also Chatterton *et al.*, Chapter 25 in this volume). Clearly, in all PAR projects, researchers and participants can make decisions about what methods are suitable: for example, collective and visual approaches (Kesby *et al.* 2005; see also Alexander *et al.,* Chapter 14 in this volume) might not always be appropriate in work with disabled people. Nevertheless, while the needs of our target group and the sensitivity of our topics necessitated the use of more conventional and individualised methods of data collection that made collaborative interaction between participants predominantly virtual, this did not prevent our approach from being participatory (see also Kindon *et al.*, Chapter 2 in this volume). We found ways to integrate participatory elements into our research design, data collection and dissemination. We shared control of the research with the participants, engaged in a genuinely collaborative process of reflection and learning, and helped instigate positive long-term change in the lives of the women involved.

## Notes

1 Centre for Independent Living, Ottawa; Lothian Centre for Independent Living; Glasgow Centre for Independent Living; Lothian Coalition of Disabled People; Access Ability, Lothian. The latter two organisations no longer exist.

# 12 Working with migrant communities

## Collaborating with the Kalayaan Centre in Vancouver, Canada

*Geraldine Pratt, in collaboration with the Philippine Women Centre of BC and* Ugnayan ng Kabataang Pilipino sa Canada/*Filipino-Canadian Youth Alliance*

### Introduction

It is interesting to listen to my long-time research collaborators at the Kalayaan Centre in Vancouver speak of the research partnerships that they have refused.[1] There was the case of the young playwright who 'woke up last week and realised that he was Filipino'. He was artist-in-residence at a local theatre, with the resources to train four writers to create a play on the experiences of Filipino migrant domestic workers and mail order brides. He came to the Centre asking for insight into these experiences, but was lukewarm to the possibility that there are youth at the Centre worth training as writers. They turned down the collaboration. In Charlene Sayo's words, 'You're going to take our stories and that's it? ...It's very personal for us, to have somebody who will write our stories but not really listen to us. Or take the process seriously.' In another case, a researcher studying violence against women approached the Kalayaan Centre for access to Filipino women to interview. Again, the collaboration was refused. 'For us', says Cecilia Diocson, 'there was no collaboration anyway, because collaboration means that we need to sit down and formulate what kind of project we really want to do: what kind of research? That is what we call collaboration' (Box 12.1).

These failed encounters display commonly held assumptions about who has methodological expertise, and the capacity to think and write creatively. Activist groups at the Kalayaan Centre were approached for their access to immigrants' experiences; the self-designated experts aimed to extract and retell these stories on and in their own terms. The benefits from collaboration seemed to flow in one direction: away from Filipino activists to the professionals who approached them.

As Cecilia notes, 'we want our stories to come out because that's part of the process of change.' Collecting and telling stories can empower a community by documenting marginalisation as a shared rather than individualised experience. Stories can articulate and validate experience. In Cecilia's words, 'It really helps [the storytellers'] own analysis, helps in their realisation that, "Oh yeah, it's really

**Box 12.1 Faux PAR**

I think they needed six more women to interview, and so they asked us for help finding them. The researcher was only willing to give 50 dollars to the women [for the interview]. But I said, 'How about the person co-coordinating the research? They're all volunteers and I think it's good if there is also an honorarium for this person ... And then finally she said, 'Oh, okay I'll give 200 dollars for this person to co-ordinate. But, you know, we have this person who is going to be doing the interviews.' Well I said, 'It's up to you but when we do the interview it's very thorough.' We want to get the stories of these women and we know that we can get their stories. So finally, because she was rushing and putting pressure on us, we just decided that no, we can't participate. We're just the last minute group they turn to if they cannot find any more Filipinos to interview. So there's no respect for the existing community researchers, who have done a lot of work already ... I said, 'There's no collaboration. Sorry, we can't do this.' ... For us, saying no under these circumstances is part of our assertion ... Yes, we want our stories to come out because that's part of the process of change. But ... if they're not respectful and they continue to have this kind of attitude, then it's fine, we're not in a rush [unlike the researcher on this occasion].

Source: Cecilia Diocson, Chair, National Alliance of Philippine Women in Canada, Interview: 18 May 2006

true, it's really happening". Like systemic racism is really happening.' It is for this reason that Kelly Oliver (2001) claims that the very act of testimony can undo some of the violence of racism (see also Cieri and McCauley, Chapter 17 in this volume). Racism constructs an individual as a subject without full subjectivity; as a statement about personal experience, testimony is testimony to subjectivity. But there is no guarantee that storytelling will be empowering. Reflecting on the risks of personal storytelling, Schaffer and Smith (2004) write of the danger of 'fixing [the] life and identity of the teller in victimhood' (2004: 45). If such narratives fix the tellers as victims, and those who collect and listen to the stories as 'advocates and agents', they merely reinforce existing power relations (Schaffer and Smith 2004: 111; see also hooks 1990; Mohanty 2003). At issue here is not only the kind of representations constructed but the very process of collecting a community's stories. Refusing to collaborate under conditions that reproduce existing hierarchies of knowledge production, of experts and naïve informants, is – as Cecilia states (Box 12.1) – an act of community assertion.

## Eleven years of collaboration

Set against these failed attempts at collaboration, we want to describe our experience of working together on four research projects over the last eleven years. We

*Plate 12.1* Collecting stories at the Philippine Women Centre, August 1995 (Credit: the Philippine Women Centre of BC, 1995)

offer no template for successful collaborations between a white, middle-class university researcher and marginalised migrants from the majority world. Indeed we have continuously renegotiated the nature and methods of our collaboration, depending on the topic and the stage of community organisation at the time. In the first project (Pratt 2004), we met for six day-long workshops with fifteen or so Filipina domestic workers who were all registered in a temporary work visa programme called the Live-in Caregiver Programme (LCP). It was possible to bring this number of women together because they worked a fairly regular five-day work week, with weekends off. The women's stories were mostly angry ones, told as a means of organising to change their conditions under the LCP (Plate 12.1).

In our next project (Pratt in collaboration with the Philippine Women Centre 2005), we brought the same women together eight years later to document how they were getting on after fulfilling the requirements of the LCP and settling permanently in Canada. Although the aim for collective storytelling remained, the effects of deskilling were evident; because so many women were working at multiple jobs, it was difficult to bring everyone together at one time, and individual interviews were arranged in some cases to allow a more thorough examination of the particularities of individuals' lives. The women's experiences also had begun to diverge – for many, their circumstances had not improved and, in some respects, even worsened.

The third collaboration was with *Ugnayan ng Kabataang Pilipino sa Canada*, a Filipino-Canadian youth organisation also housed at the Kalayaan Centre. Youth conducted ten focus groups with first- and second-generation youth to record stories of racism and feelings of dislocation in Canada, and efforts to regain a sense of home and belonging (Pratt in collaboration with the *Ugnayan ng Kabataang Pilipino sa Canada* 2003/4).

Our most recent project involves both the Philippine Women Centre and Ugnayan, and we are interviewing mothers and children, and some fathers, who have been separated for a long time while the mother worked in Canada under the LCP and her family remained in the Philippines. These are often sad stories, stories of not being recognised by one's own children, or bewilderment about one's mother's sudden departure. We have collected these stories through interviews with individuals or with mother and children together or, in a few instances, joint interviews or small focus groups with youth. Once again, our goal is to use the stories to draw out a collective, community story. One of the women who we interviewed ended her interview by saying, 'I think that we are the worst family'; our project is dedicated to redirecting this despair into the understanding that she has a remarkably strong, tenacious family, which has endured a terrible experience, structured by the regulations of the LCP. Because of the change in emotional tone, the more individualistic mode of collecting stories, and the focus on problems settling into Vancouver, the risks of victimisation, voyeuristic witnessing of suffering, and community stigmatisation seem even more pressing for this project relative to earlier ones, and the need for community ownership extremely important.

## Building trust and community ownership

How to build community ownership? For all of our projects, the research objectives and methods have been discussed and agreed upon collectively. For the first project, this involved a day-long discussion with the domestic workers who had agreed to participate: to identify themes and methods. When we met for two full days to discuss the agreed-upon themes, the domestic workers were enrolled as community researchers (rather than research participants), and they were asked to lead and report on the group discussions. We met three further times to code, verify and analyse the transcripts. In one such session, we literally passed the transcripts from person to person, each taking a turn to read a portion of the transcripts out loud, allowing an opportunity to verify, and expand and reflect upon what had been said. There was an understanding that the data were common property, but that any academic papers written from it – if not jointly authored – should be distributed and commented upon by the Philippine Women Centre prior to publication.

Because of the more fragmented nature of our current project (individual interviews at different times in different spaces), it has been more difficult to achieve a fully collaborative process, especially the inclusion of all participants as community researchers with the capacity to analyse – and not simply report upon – their situations. It has been a slower, more piecemeal process of developing interviewees as interviewers and theorists (see also Cahill, Chapter 21 in this volume). The project has sprawled over a number of years, and some of the youth who were first interviewed two years ago are now interviewing other youth. It has taken time to engage youth in this way (Box 12.2). As Charlene and Cecilia explained: 'They're so marginalised. You know, they've dropped out of school. Their education is really affected. Their sense of confidence and development is really robbed of them… It's common for youth to reject school and formal education' (Charlene Sayo); 'The interviews – before [the youth] did not want to do these, but now that they have a feeling of ownership, they're very assertive in going out and interviewing, and using [the interviews] in the programmes at the Centre' (Cecilia Diocson).

## Working through difference

Negotiating trust and shared ownership is an ongoing process, and can be an especial challenge for researchers collaborating with migrants or refugees. It was already the third of the day-long workshops of the first project, as I was sitting with three domestic workers, when Ruby turned to me and asked, 'How about you,

### Box 12.2 Prolonged engagement

[Our collaboration] has been a long process. And it's been nurtured, and we can see the long-term impacts. When we first did the project about Filipino youth, in particular, it wasn't like you just left and that was it. There was a lot more there that you wanted to explore, which was really important. And so, moving on from that project, we've looked at the issue of family separation and impacts on the youth. So there was that follow up, or at least continuity. And I think for the youth that were involved in that, they could see that…. You know, you have to understand that when you're bringing in these youth who don't even want to be here in Canada anymore because there are no opportunities for them – they're criminalised already – there's no trust. You know, they have a hard time trusting people. So when they see that their stories and their experiences are being taken seriously, and that they themselves can also develop from it, then, of course, there's really that sense of ownership and also that sense… I guess it's a better relationship knowing that their stories aren't being used to further your career or whatever. But they're really taken seriously… I know for the youth, for some of the younger ones, that when they see that their names, their stories, are being published, of course, for them it's like, wow, they're being validated. But they know they can also do it themselves. I think that's a big, huge step. So knowing that there's always that benefit there of education and that process of development.

Source: Charlene Sayo, *Ugnayan ng Kabataang Pilipino sa Canada*, Interview: 18 May 2006

Gerry, do you have a nanny?' I answered 'No', and Ruby noted, 'That's a personal question'. Marlyn chimed in, 'Do you want a nanny? Okay, take me as your nanny'. I explained that I sent my child to day-care. Ana ventured, 'She will be one of the good employers'. Scrutinising me, Ruby commented, 'You can see in her facial expression. My employer is good but a little bit Tupperware [plastic]'. I redirected the conversation by asking: 'What would be a good employer?' This is a telling snippet of conversation that disrupts any seamless story that we might wish to tell about the fully collaborative nature of the first project. My difference in status and economic standing was marked in a number of ways: in a context in which these domestic workers were exposing intimate details of their lives, they were uncertain whether it is appropriate to ask personal details about mine (and, indeed, I gave few), and there was explicit recognition that our relationship could easily slip from being research collaborators to that of antagonistic employer–employees.

But rather than focusing only on the ongoing challenges of negotiating across cultural, social and economic differences, we want to emphasise the productivity of these differences. Indeed, recognising this productivity is one means of working with – rather than attempting to overcome – difference. The benefits of cultural difference to the researcher are perhaps most obvious. The breadth of transnational cultural, historical and political understanding among the community researchers at the Kalayaan Centre, their transnational networks within the Filipino community, and their capacity to carry out interviews and focus groups in Tagalog have contributed immeasurably to the richness of the oral testimony data and depth of analysis.

The first time I worked with Ugnayan, and we met to review a draft manuscript that I had produced from the focus group transcripts, they told me to 'deepen' my analysis (cf. Cahill, Chapter 21 in this volume). This was an interesting and important moment that productively unsettled my authority as a university researcher and theorist. Ugnayan members rightly insisted that their marginalisation be framed, not only within Canadian society, but also within a much longer history of forced migration from the Philippines. This is an extremely important point: migrants inhabit many worlds, only some of which will be visible to the researcher in the minority world. Collaboration is one means of bringing the complexity of geographical experience into view. Working with activists has also ensured that the research circulates beyond the academy: through press releases, community forums, a weekly community radio programme, and various representations to government (Plate 12.2).

The benefits of collaboration run in both directions. Because of their perceived neutrality and professionalism, academics can gain access to government data or interviewees unavailable to community activists (see also Routledge 2002). For example, we have been able to establish that Filipino youth are dropping out of Vancouver high schools at very high rates, relative to other youth, using a Ministry of Education data set which tracks year by year every youth within the British Columbia school system. The BC Ministry of Education closely vets each proposed statistical analysis, and researchers must sign an agreement to obtain

*Plate 12.2* Presenting our research at the Kalayaan Centre, Spring 2006 (Credit: the Philippine Women Centre of BC, 2006)

Ministry approval before publishing or presenting material that draws upon this data.[2] Though it is speculation that the proposals of UBC academics and community activists might be evaluated differently, I was told informally that passage through the evaluation process is eased if the project has been reviewed by the University's Ethics Review Board (see also Manzo and Brightbill, Chapter 5 in this volume). Certainly any investigator requires the financial resources to pay for the data analysis, which can only be done by Edudata Canada data analysts.

Beyond access to certain kinds of resources, association with an academic can authorise ongoing community research – even within the Filipino community (Box 12.3). Along with corroborating and legitimating existing community research, our collaboration has built the Centre's capacity to generate funding from government agencies to do more of their own community research. The latest and largest is a three-year project, which began in spring 2006, funded by the Department of Canadian Heritage and carried out by the National Alliance of Philippine Women in Canada, to examine factors leading to the economic and social marginalisation of Filipino communities in Canada, and to strategise towards their fuller participation. It was a long struggle to secure this funding, in part because the government considered this a large research project for a community group to carry out on its own. The Alliance's long track record of collaborating with many different university researchers no doubt strengthened their case for independent funding.

The federal government was also loath to fund this community research project because its focus on a single ethnic group challenges existing governmental understandings of multiculturalism. In Cecilia Diocson's words, 'They really have a

**Box 12.3 Academic legitimisation**

Gerry, a classic example is when we say that Filipino youth have … one of the highest dropout rates among young people in the Lower Mainland. It's just a statement if it's not backed up by an academic researcher. The credibility is not really that strong within the community. Unless we show them, 'Look, Dan Hiebert, Gerry Pratt, these are their findings' and all that stuff. Then even the community is surprised … now it's being backed up by this community research from the academic. Then that becomes a very powerful tool, and suddenly people start using it, and it just spreads out … The members of the [broader] community can see that here's the Kalayaan Centre. The credibility of the Kalayaan Centre is also bolstered by the fact that whatever we say at the Centre is backed up by very strong academic research.

Source: Emanual Sayo, British Columbia Committee for Human Rights in the Philippines, interview: 18 May 2006

different concept of democracy because they think that having other people there outside of the Filipino community is the essence of democracy'. But 'What's the essence of democracy?' Cecilia asked, and then, by way of an answer, described the first weekend conference that launched the project (Box 12.4).

Rethinking and revising the process of doing research is one important aspect of this larger project of rethinking democracy. The paradigm of Participatory Action Research provides no easy methodological blueprint for doing this. Indeed, we have learnt that the research process has to be responsive to the circumstances being researched, and that it can take time (sometimes a very long time) to build the trust, skills and community enthusiasm necessary for this type of collaboration (see also Hume-Cook *et al.*, Chapter 19 in this volume). Rather than a set of rules or techniques, what this research tradition provides is a shared commitment to

**Box 12.4 Understanding democracy**

You know the whole weekend people are talking, participating, and putting their ideas forward. That's democracy in itself. So I was talking to [the federal government project officer who oversees their project and attended the conference], and I said, 'You see it's very participatory, it's very democratic.' I mean, really, are you the only people who understand what democracy is? I think we really need to examine and re-examine our… understanding and practice of democracy.

Source: Cecilia Diocson, interview: 18 May 2006

fundamentally disrupt conventional hierarchies of knowledge production: who decides on the questions to ask, how to ask them, and how to theorise the world. Given the multiple hierarchical differences that separate a white middle-class Canadian researcher from migrant Filipino women, this has been foundational to our research practice.

## Notes

1 This paper was developed from a conversation with Cecilia Diocson (Chair, National Alliance of Philippine Women in Canada), Charlene Sayo (*Ugnayan ng Kabataang Pilipino sa Canada*/Filipino Canadian Youth Alliance) and Emanual Sayo (BC Committee for Human Rights in the Philippines) on 18 May 2006. All of these organisations are housed at the *Kalayaan* Centre in Vancouver. This particular means of collaborative writing was the one chosen by the activists involved.

2 I received a letter from the Ministry two weeks after making an oral presentation without obtaining this approval: 'It has come to our attention…'

# 13 Peer research with youth

## Negotiating (sub)cultural capital, place and participation in Aotearoa/ New Zealand

*Jane Higgins, Karen Nairn and Judith Sligo*

## Introduction

**Scene 1** In a high school assembly hall, two young researchers, working with a teacher, organise a special assembly in order to recruit participants for a university research project on student rights in schools. Although these young researchers lack the authority to call such an assembly, they have the ability to liaise with a teacher who can do this.

**Scene 2** In a busy pedestrian mall in a different city centre, two young researchers approach other young people as they go by to ask them what they think of City Council initiatives to involve youth in the life of the city. The researchers are quick to establish rapport with many potential interviewees because each is able to 'read' the other in terms of dress and hairstyle, ways of speaking and general manner, all of which convey important information about personal tastes, particularly musical tastes, relating to youth cultures and subcultures.

As different as they may seem, these research scenes share two important characteristics. First, and most obviously, they both involve young researchers working with young research participants. Secondly, and less obviously, in each case attempts by the researchers to engage with potential participants draw on a particular congruence between place on the one hand and, on the other, the abilities of these researchers to trade on their own subcultural and cultural capital. In this chapter we explore how young people, as researchers, are able to make connections with other young people in the research process, and through those connections to enhance this process. We emphasise particularly the significance of place in the playing out of these research relationships.

## Youth researching youth

In recent years, the practice of conducting youth research with (rather than simply on or about) young people has been an important methodological development in youth studies (for example, see Broad and Saunders 1998; Clark and Moss 1996; Schensul *et al.* 2004). This has been driven by developments such as the United Nations Convention on the Rights of the Child (UNCROC) Article 12 establishing the right of young people to participate in decisions and activities that affect them.

There has also been strong momentum in the social sciences towards methods involving peer research, through building research teams that include at least some members of the group that is being researched (Alder and Sandor 1990; Kelly 1993; Oldfather 1995). Peer research addresses some of the key concerns of Participatory Action Research (PAR) in that it recognises that individuals within any community being researched are themselves competent agents, capable of participating in research on a variety of levels, including as researchers.

The advantages of this approach in relation to youth peer researchers are clear – or at least they seem to be. First, in order to hear and honour young people's voices in research it makes sense to engage with them on as many levels as possible, not just as participants but also in project design, implementation, analysis and dissemination (Fielding 2004; France 2004; Garcia *et al.* 1995). Secondly, power imbalances inevitably inflect research relationships between adult researchers and young participants. Involving peer researchers offers an opportunity to remedy those imbalances to a greater or lesser extent (Christensen 2004; Cook-Sather 2002). Finally, adult researchers are often keenly aware that a wide range of potential youth participants lies out of reach because of age and socio-cultural distance. With peer researchers those distances may be reduced, making access less of a difficulty (Broad and Saunders 1998; Smith *et al.* 2002).

These considerations provide powerful arguments for youth peer research. It does not follow, however, that involving young people in this way is an unalloyed 'good' requiring no further reflection. Our own experience of working with peer researchers on projects involving youth has convinced us of the importance of reflexivity in this area. The reasons for this are twofold: we wish to avoid the tokenism of a 'just add young people and stir' approach, and we are keen to better understand the contribution that peer researchers can make to a research team.

## Trading on cultural and subcultural capital

We have developed elsewhere one possible theoretical approach to youth peer research (Nairn *et al.* forthcoming). Here we explore this approach in relation to three projects in which we have worked with young people in this way (see Box 13.1). We use these case studies to look first at the nature of the contribution that

### Box 13.1 Research questions on three projects

Our three projects involving youth researchers asked these research questions:

1. How do high school students perceive their rights at school?
2. How do young people view local government initiatives aimed at facilitating youth involvement in city life?
3. How do young people articulate identities at the child/adult border?

Source: Jane Higgins, Karen Nairn and Judith Sligo

peer researchers can make to a research team. We then examine some of the ways in which the location of the research influences the research relationship between peer researcher and participant, and consequently the nature of the data collected.

Each project involved young people interviewing other young people. Adult researchers were also involved, variously, in interviews and focus groups.

To begin, we return to our two earlier examples. In both the student rights project and the City Council project, peer researchers made use of forms of knowledge that they possessed about school cultures and youth cultures respectively, to help them make connections with research participants.

When Naomi (who is Chinese) and Kathy (who is *Pākehā*)[1] negotiated with a teacher in their school to call a special assembly in order to recruit participants to the rights project, they were using their understanding of how their school operated, from its lines of authority to the systems by which students could be accessed and information relayed to them. Naomi indicated this when she explained that any researcher seeking student participants should 'have a sense of how the school runs and how assemblies are organised and how it is easier to get access to people' (Debriefing interview, 12 June 2001).

Understandings such as this constitute forms of cultural capital, that is, forms of knowledge and dispositions of mind and body that enable the bearer to operate with ease within the dominant (usually middle-class) culture of his or her social space (Bourdieu 2004). Being able to put such understandings to work is a form of 'trading on' that cultural capital. Kathy and Naomi were able to do this because their cultural capital included the appropriate knowledge and skills to work easily with teachers and within school processes. Speaking of the teacher who helped them, one said:

> I mean, she organised the assembly and, you know usually we wouldn't be able to do something like that ourselves because it's run by the teachers so it just helped us out a bit, yeah. You know, sort of someone with a bit more authority just to get us out there.
>
> Debriefing interview with Naomi and Kathy, 12 June 2001

University researchers are well used to deploying cultural capital: our educational qualifications are a classic sign of this kind of capital. We therefore find it fairly easy to relate to youth researchers who also possess (middle-class) cultural capital. It is a straightforward matter to incorporate them into research teams because they seem like youthful versions of ourselves. But to construct them in this way, simply as less qualified or less competent versions of ourselves, is to judge their performance by our own standards. In doing this we fail to understand the contribution that they can make within a research team. Kathy and Naomi used their cultural capital to set up the assembly, but when it came to interviewing participants they also recognised that the cultural capital that was useful for communicating with teachers was not useful for communicating with younger students:

> We hear a lot of them you know name calling and stuff like that and we think it's them being unkind but I had quite a few third formers [12 to 13-year-old students] that actually said they just do that for fun … to their friends and stuff, but we don't realise that … *not being a part of them we don't realize that.*
>
> Debriefing interview with Naomi and Kathy, 12 June 2001 [emphasis added]

This is a key insight. In their name-calling, the third-formers were arguably developing a form of *subcultural* capital to which senior students were not privy. By subcultural capital we mean forms of knowledge and dispositions of mind and body that are not officially sanctioned and that may be developed precisely in opposition to the 'mainstream'. The term grew out of Thornton's (1995) work on how young people exhibit ways of being 'in the know' about what is and is not currently popular in language, fashion, dance and music. We use it here to explore the capacity of peer researchers to connect with other young people.

In the City Council project, Rachel (who is *Pākehā*) and Moana (who is *Māori*) worked as peer researchers; they possessed considerable cultural capital as they were trained youth workers, but this was not what helped them to establish rapport with young people on the street. Rather it was their knowledge of current youth subcultures and their own manner and appearance that put the young people they approached at their ease. This would surely not have happened had older researchers approached them.

In debriefing interviews these two researchers provided insights into how their own (different) forms of subcultural capital helped them to approach different groups. Moana observed:

> I think also we actually walked round and at one point we were saying, 'Oh, shall we do these guys?' I think it was the 'metalers' and Rach goes, 'Oh, let's do these guys'. But it was a bit out of my comfort zone. But we still did it. And I said, 'Oh, well let's do these guys?' And she says, 'Oh, it's a bit out of my comfort zone', but we did it.'
>
> Debriefing interview with Rachel and Moana, 5 December 2001

Rachel commented:

> It's like the – I mean it is a bit stereotyping – the way people look. But the different youth cultures. Like, the 'metaler' guys, because I myself have had experience in that scene and that culture, then I'm, like, sweet you know, they don't bother me at all. But those two girls, with the far out hair, they were just like, the kind of people – and I don't like to say this – but they are the kind of people that I judge. That youth culture that is real alternative – way out.
>
> Debriefing interview with Rachel and Moana, 5 December 2001

Moana spoke of why they were able to engage with participants:

> I think it's their perception of us, as well, that comes into play, big time. … And I think the fact that we both look so different as well. They can probably relate to one of us more than the other.
>
> Debriefing interview with Rachel and Moana, 5 December 2001

These examples suggest that the subcultural capital that young people possess is likely to be an asset on any team researching youth. This is not to say that all young people possess the same kinds of subcultural capital or the same balance of cultural and subcultural capital. It is clear, however, that research teams should recognise subcultural capital as a legitimate form of knowledge and a significant asset when working with young people.

## Subcultural capital and place

Once we recognise the significance of subcultural capital in youth research, it becomes important to consider how readily peer researchers are able to put their subcultural capital to work in the research process. What, for example, is the role of place in enabling peer researchers to connect with other young people? The case studies we have been examining suggest that it has a significant role. For example, Rachel and Moana also conduct research in schools as part of their youth work.

> We have focus groups and we invite, or rope-in, or make [laugh] students talk to us and teachers. I mean, a lot of time when we contact schools … the teachers will organise the students. And I wonder how much say the students actually have in talking to us. They just say, 'We want a certain amount of people – you're doing it'.
>
> Debriefing interview with Rachel, 5 December 2001

This is a world away from approaching young people on the street. In a focus group held in school time on school grounds, rules apply with respect to power relationships. Teachers act as gatekeepers and authorities. It is not hard to imagine that students might view the researchers as teacher equivalents, particularly if they have been 'roped in' to participating. The opportunities for recognition of mutual subcultural capital are likely to be constrained in this setting. And this surely has an impact on the type of data gathered. In contrast, Rachel and Moana both observed that it was relatively easy to approach potential participants in the pedestrian mall because these young people were comfortable in that place – 'just really hanging out' (debriefing interview, 5 December 2001).

> An advantage of going to them [young people hanging out] is that it is within their own element – it's in their own comfort zone and they feel comfortable

> about doing that. ... The disadvantage of them coming to us is that it's out of their comfort zone. They don't feel as comfortable expressing themselves and saying things.
>
> Debriefing interview with Moana, 5 December 2001

That the school setting is likely to privilege cultural over subcultural capital is also illustrated by other peer researchers in the student rights project. One of the male peer researchers observed that the research 'was just like school work really ... you just asked a few people some questions and wrote it down' (debriefing interview with Anthony and Graeme who are both *Pākehā*, 23 August 2001). Thus the project was viewed as an exercise in the practice of cultural capital – school work. But these boys did not have the authority of the teachers who are the exemplars of cultural capital, and this created problems.

> You'd ask them questions and they'd give you a stupid answer that was totally immature, kind of thing. ... It was just hard to get some of them to take you seriously. Coz you are not an adult.
>
> Debriefing interview with Anthony and Graeme, 23 August 2001

Had these boys been able to talk to other students in a mutually agreed setting they may have been able to relate to each other in a different way, on the basis of joint subcultural rather than cultural capital. So we are suggesting that different places privilege the exercise of different capitals and that this has implications for the data collected and the modes of participation that are possible.

In another project, about young people's school to post-school transitions, we offered peer researchers and participants the opportunity to meet in a mutually agreed space. We expected that this would allow interviewer and participant to meet in a way that set both at ease, perhaps through recognition of shared subcultural capital. We also expected that participants' homes would provide the most convenient location for this. But with a few exceptions, most participants chose to be interviewed in youth researchers' homes, university or school. These choices are significant.

The home represents the cultural capital of a family. A visitor can read a family's social class in the material detail of their home including the amount and arrangement of space, items of furniture, wall hangings and books. Similarly a teenager's own space, a bedroom for example, can be interpreted as representing subcultural capital (through images on walls, and material goods such as cellphones and clothing). In inviting a researcher home, a participant is making these relatively private spaces available for assessment. There is some risk in this in terms of whether a participant feels that their cultural and/or subcultural capital is congruent with that of their visitor.

This risk may be allayed in a non-personalised space. Maggie, one of the (*Pākehā*) peer researchers on this project, suggested that her participant might have limited what she said during the interview in order to manage the impression she gave: 'you might be worried about what the other person thinks because they are of your age' (debriefing interview, 9 February 2005). Peer research may not provide a comfortable process for participants to share information about themselves if researcher and interviewee are concerned about differences in subcultural capital, particularly different tastes and the associated hierarchies of what counts as 'cool' amongst youth. This may be one reason why so few of our peer researchers' interviewees in this project chose to be interviewed at home.

But even places where subcultural capital might be shared without risk to participants, notably peer researchers' homes, were not necessarily conducive to data gathering. Comments from two of the young male researchers, Marvin and Hamish (who are both *Pākehā*) suggest this. Marvin, who interviewed in his home, indicated that the formalities of the interview (consent form and audio-taping) overwhelmed the informality of the home environment and inhibited data collection (debriefing interview, 15 September 2003). Similarly, Hamish, who conducted his interviews at his kitchen table, reported that although both participants were friends, neither shared personal information that he himself already knew about through the context of their friendships (debriefing interview, 14 October 2003). The establishment of rapport is therefore not guaranteed in interviews between peers even in places chosen by participants.

Other peer researchers' reflections on interview spaces outside the home further illustrate how these spaces influenced the modes of participation of their interviewees. Alice (who is *Pākehā*) conducted one interview at her home and the other at school. She noted how the participant interviewed in an empty classroom took the interview more seriously than the participant interviewed at her home. Similarly, Dana (who is *Māori*) felt that the university classroom gave her interview a professional quality: 'I definitely felt professional, like a space between us' (debriefing interview, 6 January 2005). Dana thus traded on her cultural capital as a research interviewer in a space that was congruent with this form of capital. This example illustrates how youth researchers' own interpretations of their role is also influential in shaping what capitals they might deploy in their research encounters.

## Conclusion

Youth researching youth is an important goal to pursue in participatory research endeavours. Our experience across three projects suggests that the inclusion of young people in a research team can enhance research in ways not possible when all the researchers are adults. Youth researchers are able to employ a mix of cultural and subcultural capital that adult researchers are highly unlikely to possess, and to use this mix in useful ways to recruit and establish rapport with young research participants.

In this chapter, we have argued that the space in which the research takes place plays a key role in shaping the use of different forms of (sub)cultural capital by peer researchers. Street, school and home all provide very different contexts for the deployment of these forms of capital and this in its turn influences the modes of participation that are possible for participants, as well as the types of data generated.

Engagement between peer researcher and research participant will, of course, vary according to cross-cultural and cross-national differences as well as across subcultural lines. These differences will in turn be played out in particular ways in different spaces. This means that research teams will need to adapt peer research procedures accordingly. An awareness of how cultural and subcultural capitals are displayed in space will provide a useful theoretical tool for doing this and for understanding why particular participatory research strategies operate the way they do in particular spaces.

## Notes

1 *Pākehā* is a highly contested term usually used to refer to the descendents of white colonial settlers to *Aotearoa*/New Zealand.

# 14 Participatory diagramming

## A critical view from North East England

*Catherine Alexander, Natalie Beale, Mike Kesby, Sara Kindon, Julia McMillan, Rachel Pain and Friederike Ziegler*

### Introduction

Diagramming is the product of a long participatory evolution within various development projects in the majority world, and today widely used elsewhere (see Kindon *et al.* Chapter 2 in this volume). Using graphic and/or tactile materials to create visual representations that express participants' ideas and understandings, Participatory Diagramming is often used with groups to draw out issues and galvanise further discussion, analysis and action.

Diagramming *per se*, however, is not without its challenges and does not necessarily constitute Participatory Action Research (PAR). We explore some of these challenges and their associated critiques through reflexive discussion of researcher experiences. In the spirit of diagramming's collaborative tradition, we have produced this chapter as a collective, refreshing the accounts of those with considerable experience (Kesby 2000b; Kesby *et al.* 2005; Kindon 1995a; 2003; Pain and Francis 2003) with those of emerging researchers working in North East England (see Boxes 14.1–14.4).

### Understanding *participatory* diagramming

We wish to locate this discussion within the distinction between participatory approaches and participatory techniques (Kesby *et al.* 2005; see also Kindon *et al.* Chapter 1 and Kindon *et al.* Chapter 2, in this volume). Generating diagrams often has participative aspects, but diagramming is easily and often deployed in isolation from wider participatory processes and procedures (i.e. engaging participants only in data collection). There is nothing necessarily wrong with such deployment, but it should be described honestly. Conversely, when less participatory techniques (for example, interviewing, questionnaires, GIS) are deployed within a participatory framework, their politics and effects are completely changed (see for example Opondo *et al.* Chapter 10, Higgins *et al.* Chapter 13 and Elwood *et al.* Chapter 20, in this volume). Box 14.1 describes the use of diagramming to kick-start a participatory process.

## Box 14.1 Diagramming as means to participatory research design

When planning to research children's journeys to school in Gateshead, UK, I imagined using questionnaires, photography and personal travel diaries. However, I wanted my project to be participatory, so I invited 54 boys and girls aged 10–11 years to use diagramming to choose the methods they thought would best help them describe their journeys. The first important step was to let participants chose their own groups and location for work.

Using participatory diagramming for research design was a revelation (see Plate 14.1). Using diagrams and stickers to brainstorm possible options showed that *none* of them wanted to use my preferred methods! There was some interest in the use of photography, but my cameras weren't considered advanced enough! Diagramming revealed a desire among a majority of the boys to write rap songs (using drums, piano, dress-up clothing and role play) to convey their journey experiences. The two mixed-sex groups preferred group discussions and interviewing, while poetry, mapping, film making, drawing, Powerpoint presentations and play-scripting and performing were favoured by the girl-only groups.

*Plate 14.1* Favoured methods of all children (Credit: Julia McMillan 2006)

*Box 14.1 continued*

My experience was very instructive: first, the early deployment of diagramming techniques enabled my research to become more participatory. Second, early diagramming helped the children create a varied set of 'alternative research methods' that enriched the project and increased participants' sense of ownership over it. Third, diagramming allowed children to develop methods appropriate to their own skills and learning styles. Analytical thinkers preferred mapping, whilst movement, role play, song and poetry appealed to the more intuitive and creative individuals. Teachers noted a marked and unexpected improvement in motivation, behaviour and performance of some children. Finally, when techniques are deployed through a participatory framework, research methodologies can become resources with which children (or adults) can affect their agency (Kesby 2007a). Not only did children assert their opinions and initiative in a way that the researcher and teachers had not anticipated, they also simultaneously developed the very core skills and competencies (creativity, problem solving, leadership, self-esteem, confidence, co-operation, negotiation, responsibility, decision making, reflection and critical appraisal) that schools are desperate to promote.

Pursuing a participatory *approach* (open to being guided by participants) rather than being wedded to diagramming as a particular *technique*, I had enhanced my PhD project, and facilitated enjoyable, fulfilling and empowering learning among participants.

Source: Julia McMillan

In common with other group discussion techniques, researchers facilitate the contributions of participants, but several features alter the nature and outcomes of sessions. Firstly, diagramming exploits *visual* as well as oral methods to express, organise, represent and disseminate information. There are many formats to adapt, limited only by one's imagination, for example maps, transects, sketches, cartoons, pie charts, flow diagrams and matrices (see Chambers 1997; 2002; Kesby 2000b; Kesby *et al.* 2005; Kindon 1998; Mikkelson 1995; PLA Notes 1988–present).

Secondly, diagramming can be more *participatory*. In focus groups, researchers aim to facilitate discussion, but often end up the centre of attention as they manage a series of solo oral contributions. When a diagram becomes the focus of attention, several people can contribute at once. As they engage with the research tools very directly, the balance of control, knowledge production and analysis shifts in their favour. In global south contexts, diagrams have often been created using locally available materials: seeds, bottle tops, pebbles, sticks, straw, household objects or representations drawn in the soil (Chambers 1997). Flip chart paper and marker pens are now often used, with post-it notes and coloured sticky dots added to 'layer' or analyse diagrams (see Plates 14.1–14.4); techniques and materials should feel comfortable to those handling them. Diagrams are flexible, so they can be changed and adapted as ideas and understandings develop.

Thirdly, diagramming can be more *inclusive*. Using tactile and visual skills can facilitate the contributions of individuals who are less confident or eloquent, who have limited literacy or do not speak the same language, or who for cultural reasons may be in the habit of taking a back seat. Diagramming can break the ice and make people feel more positive about engaging in the research. People can make significant contributions without necessarily having to verbally articulate their opinions, and look at the diagram rather than engage in face-to-face verbal exchange. Diagramming has become popular in PAR with socially marginalised groups, but choosing a technique cannot guarantee ethical practice or a positive response. Flexible deployment and continual reflexivity are vital (see Box 14.2).

Finally, participants actively engage in data analysis, not just collection (see also Cahill, Chapter 21 in this volume). A wide range of issues can be covered relatively

**Box 14.2 Different responses to diagramming**

Diagramming is not applicable, nor appropriate, to every research context. I have learned to give participants a choice of techniques at the start of a session, and to strive to build relationships with them to achieve inclusive and mutually beneficial working. The techniques themselves cannot create a participatory process.

My PhD concerns 'local networks of rumour'; how and why young people talk about crime. The pilot study, in a deprived neighbourhood of Newcastle upon Tyne, UK, recruited three groups from local youth clubs. In an effort to engage them in a collaborative research process, I experimented with the following diagrams:

*Spider diagrams* a brainstorming exercise to get participants thinking about the neighbourhood as 'safe' or 'risky';

*Simple two columned tables* participants drew up lists of 'positives' and 'negatives' of growing up in their area (Plate 14.2);

*'Creativity through stationery'* participants were encouraged to be creative with marker pens, flip chart paper and post-it notes during the sessions, to feel free to collaborate in any way they preferred.

These techniques worked well with the first group, a boys' football team, but less well with the second group of young people with learning difficulties. This group had short attention spans, seemed disinterested in diagramming and were boisterous. Although they eventually settled down and chatted informally with me, they remained uncomfortable about annotating the diagrams themselves, preferring me to write (Plate 14.2). I realised I should have tailored the diagrams to avoid *written* annotations. The group nevertheless enjoyed seeing their own contributions, pointing them out when friends joined the session.

*Box 14.2 continued*

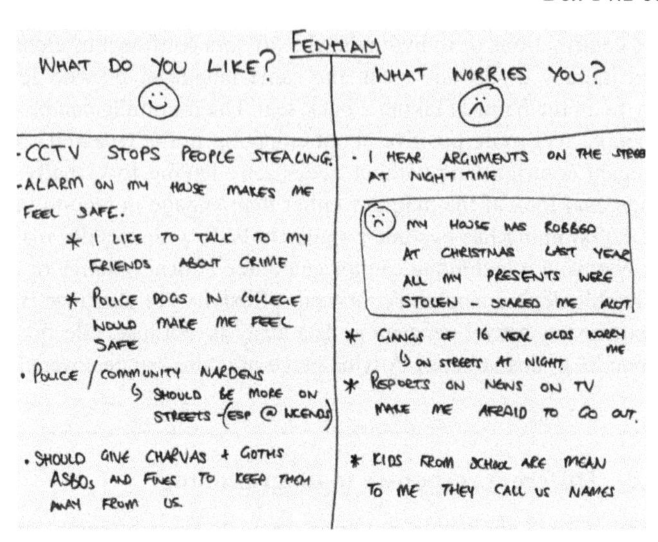


*Plate 14.2* Different responses to diagramming (Credit: Catherine Alexander 2006)

Diagramming failed completely with a third group of girls involved in the youth justice system. They saw the diagrams as patronising and/or intimidating, and resisted my strategies (ice breaker exercises and prompt cards) to engage them. After a break, we opted to chat about crime in the area, and almost immediately they discussed numerous rumours. Later, the girls suggested other techniques they would have preferred such as cameras, diaries and questionnaires (cf Box 14.1). Although diagramming was a disaster with this group, they went on to become the most committed to participating in the long-term and collaborative stages of my research.

Source: Catherine Alexander

quickly, and difficult subjects tackled more readily. Once out in the open in a diagram, these can be more fully addressed. Diagrams are typically used in sequence (for example, brainstorming problems or opportunities, identifying priorities, exploring causes and impacts in detail, then suggesting solutions or actions to address them; see Kesby *et al.* 2005). For these reasons, a visual product is not a finished result that 'speaks for itself'. Instead diagrams are interviewed (participants interrogate the meanings behind each element) and this process often provides the richest material. The depth and quality of the researcher's analysis is improved by checking it against the interpretation of participants ('member checking'). This collaborative analysis contrasts with conventional research, where individual data are extracted for analysis elsewhere. Diagramming begins to transform into participatory research when it facilitates participants' own learning, self-reflection and action through this process (see also Stuttaford and Coe, Chapter 22 in this volume).

## A critical view

It is important not to idealise the potential of diagramming. Participative engagement, inclusivity and social levelling are not easy to facilitate. People and places always exert their own influences over the research process, and diagramming will not always 'work'. We cannot predict who will respond to visual methods, and who would rather just sit and talk (see Box 14.2).

In recent years, diagramming has been subject to broader criticism as part of the wider reappraisal of participatory research and development (see Kesby *et al.* Chapter 3, Pain *et al.* Chapter 4 and Chatterton *et al.*, Chapter 25 in this volume). Critiques have been levelled at the idea of producing 'consensus', and the way that visual forms may become presented as 'the community' view. There is a need to guard against the 'tyranny of the group' (Cooke and Kothari 2001b), and it is useful to combine group sessions with other methods that forefront individual perspectives (see Box 14.3). Diagramming is not a neutral technology (Cleaver 2001), but already laden with our perspectives, values and priorities (Kothari 2001;

### Box 14.3 Participatory mapping with older people

I used participatory mapping as part of the Getting Around project, to elicit older people's perceptions of their neighbourhoods, trace their daily mobility and investigate physical and social exclusion in County Durham, UK. I was interested to explore barriers to mobility (for example, access to local facilities and services) and aspects of neighbourhoods they liked or disliked.

**Sketch maps or base maps?**

Often in PAR, participants create community maps from scratch. With older people, this may initially prove difficult; some have limited motor skills, lack confidence with tactile tasks, or fear what they see as 'school room' tests of knowledge. I found it better to begin with a printed base map showing roads and buildings on to which participants could draw. I also observed that when using very detailed printed maps, participants spent considerable time identifying *exact* locations, and were more cautious about adding their own features and perceptions. When using base maps containing less information, however; more discussion was generated as to the *relative* position of important features. Individuals felt comfortable adding to them without worrying about spatial accuracy, and the process became more participatory. It was also useful to augment the abstract map information with participants' photographs to provide 'real world' illustrations (Plate14.3).

*Box 14.3 continued*

*Plate 14.3* A sketch map with photographs taken by participants (Credit: Friederike Ziegler 2006)

**Scale**

When using base maps, researchers inevitably set the scale at which participants work. In my project, large scale maps enabled participants to mark individual features such as houses and footpaths. The printed maps also had the advantage of being to scale, which facilitated further spatial analysis using Geographic Information Systems. However, base maps inevitably limited the area of coverage, and participants (particularly those in isolated locations) often made hand-drawn additions at the maps' edges. While these lacked accurate scale, they offered more flexibility, and showed that once they lost their inhibitions, older people were able to generate their own maps.

**Collective and individual mapping**

The collective production of community maps proved a powerful way for participants to collate and distil individual mental maps. Most groups worked towards reaching a consensus before committing information to paper. However, as with all collective methods, there was a danger that minority views could get subsumed. I had to ensure careful recording of group discussion during collective diagramming to capture these. I also asked participants to produce individual maps before or after a collective map had been produced, to interrogate its universality.

Source: Friederike Ziegler

Mohan 2001). For example, if not interviewed systematically, diagrams can tend to condense the complex multiplicities of lived realities into rather linear and formalised representations (see Kothari 2001).

While diagramming might be inclusive of those who participate, we must not forget those who do not. The public nature of the technique may systematically exclude certain groups (see Guijt and Shah 1998; Pain and Francis 2003). More perniciously, the claim of inclusivity may act to exclude and delegitimate those who decline to take part (Cleaver 2001; Kothari 2001). We should carefully consider how we are deploying diagramming: research sessions that 'feel good' for researchers and participants may seem an especially deceptive power play where there is a failure to deliver any other outcomes.

Like all PAR practices, diagramming is contextually sensitive, and contingent on

## Box 14.4 Contextual effects on participatory diagramming encounters

However carefully diagramming is selected and implemented, results are always grounded in, and influenced by, the ways each research encounter has been negotiated and performed: by gatekeepers, participants, researchers

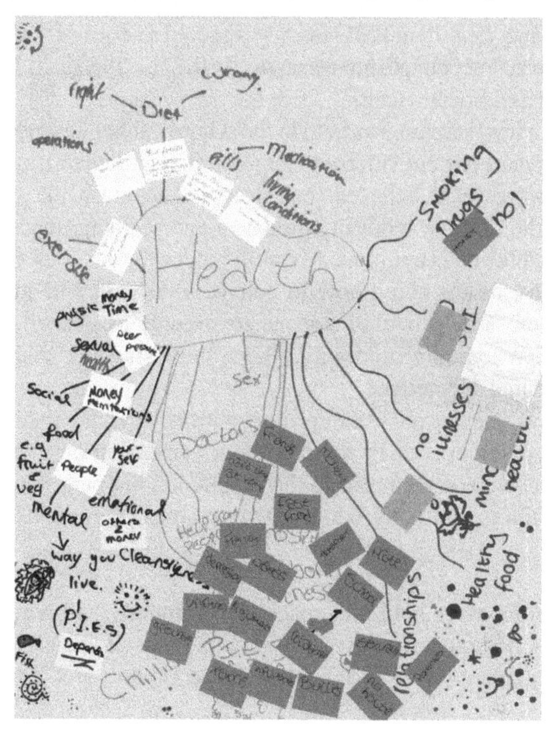


*Plate 14.4* A multi-stage diagram on meanings of health (Credit: Natalie Beale 2006)

*Box 14.4 continued*

researchers and settings and through the wider social and cultural networks in which these operate. My pilot fieldwork in County Durham, UK, explored young people's perceptions of health and risks to health. Using a multi-stage diagramming exercise (see Figure 14.4) students were asked to brainstorm meanings of health, and then add post-it notes showing risks and barriers. Setting played a significant role in shaping sessions and diagrams. Although I tried to ensure that participants felt comfortable, the institutional culture and ethos that permeated allocated spaces affected the dynamics.

- In 'Blakely' school we were allocated the library because there was plenty of space and the librarians could supervise us from a distance. However, behaviour in this space was governed by institutional expectations and reactions to these. I felt we had to be quiet and well behaved, whereas the students seemed interested in having a 'doss' lesson. Blakely's authoritarian ethos and the library context pushed me into adopting an authoritarian role. This atmosphere combined with the fact that students had recently done 'health' in PSE lessons triggered a series of 'set' responses; for example, recycling the acronym 'PIES' – Physical, Intellectual, Emotional and Social Health.
- The 'Highview' session was held in the school hall which enabled participants to spread out but suffered from competing demands on the space. Students were 'well behaved' but struggled to grasp the participatory nature of the session, requiring guidance and seeming unaccustomed to being asked for their opinions. My preferences for the session apparently went against norms about how to behave in school, and students were unable to cross these within a space so strongly associated with school behaviour. When compared with others, these students' diagrams had a more regulated appearance.
- A session at 'Netherton' youth project with boys with behavioural problems took place in a small room without ventilation on a very hot day. Participants were encouraged to dip in and out, but persistent comments about stuffiness and body odour disrupted the session and limited data collection.

Other sessions were held in a spacious community hall which could be rearranged as participants chose. This was the most conducive venue and, although participants were in school uniform, I didn't feel it was 'school-like'.

Source: Natalie Beale

participants' understanding of the research context (Cornwall 2004a; 2004b; Pain and Kindon 2007). Place provides a material setting, which may be more or less friendly to the materials we use to construct diagrams: loose natural materials used outside can blow away, flip charts can get soaked, crowded rooms get stuffy. Settings also enable or constrain the physical spacing of individuals, communication and collaboration. We know that identity and behaviour are made in and through space, so need to be sensitive to the social and political ambiences embedded in the locations that we choose, or are allocated, for our research, and how they might affect participants (see Box 14.4).

It is too easy to imagine 'community' as the scale at which diagramming focuses, but as Box 14.3 illustrated, we should beware of imposing a particular scale on a phenomenon, either in the way we ask participants to imagine it or the diagrammatic mechanisms through which we invite them to express it (Mohan and Stokke 2000) Fortunately, participants are capable of resisting our impositions, and diagramming and other participatory methods facilitate that resistance better than most. But we need to be open enough to hear or see this resistance for what it is, and think about what it might tell us (see also Kesby *et al.*, Chapter 3 in this volume).

## Closing remarks

However difficult the sites of research may be, they always hold the possibility of change (Kesby 2005; 2007a). Participatory processes may identify the very relationships between power and space that we fear will restrain them, and may be used consciously to engage with and alter them, with or without the tacit knowledge of the researcher. On the other hand they may leave these orders unchanged, or changed for the worse. All experiences of Participatory Diagramming discussed in this chapter are embedded in and constitutive of social relations, material sites and spaces (see Kesby 2005), and a sensitivity to these, we would argue, is vital for successful PAR.

# 15 Participatory cartographies

## Reflections from research performances in Fiji and Tanzania

*Eleanor Sanderson, with Holy Family Settlement Research Team, Ruth Newport, and Umaki Research Participants*[1]

### Collaboratively creating cartographies?

Cartographies connect people and places through the powerful representational practices of mapping. However, the way particular people become cartographically connected to particular places in mapping practices have not necessarily involved those people in determining the character of the connections being made. Maps do not represent relationships between people and places in a value neutral way (see Robinson 1994). Instead, the way maps are conceived, as well as created, reflects the cartographic perception of the map-makers and the power relationships informing that perception.

Cartography specifically refers to the conception, as well as the methodology of mapping. The growing awareness of the power dynamics inherent in this conception has lead to increasing critical discussion surrounding cartography (see Crampton and Krygier 2006). It is therefore important for 'less powerful people' to access, and potentially re-conceive, mapping practices (see also St. Martin and Hall-Arber, Chapter 7, and Gavin *et al.*, Chapter 8 in this volume).

Participatory Action Research (PAR) practices facilitate access to the powerful dynamics of mapping, but they also face a serious challenge. This is a challenge to collaboratively work with research participants in a process that does not confine cartographic perception to the 'external' researcher alone (also see Elwood *et al.*, Chapter 20 in this volume). Addressing this challenge is the aim of participatory cartographies. Through this chapter I reflect, with my research participants, on some of the practical ways that we tried to reach this aim and create our own participatory cartographies. First, it is important to explain participatory cartography in more detail and the methodological framework involved.

### A suggested outline for participatory cartography

To illustrate what I mean by participatory involvement in the conception, as well as the methodology of mapping, I will relate an experience that was shared with me by a range of development practitioners involved in a social mapping exercise. These individuals worked together to evaluate the impact of a women focused development Non Government Organisation (NGO) based in India. One of their evaluation activities involved a picture based mapping exercise that showed the different relationships and influences upon particular women before and after their work with the NGO. This 'paper and pen' approach to mapping changed into a performed mapping activity as a result of reflections between these women and NGO staff, and the perception that a role play performance would better facilitate the participation of all involved. In particular, the women reflected that despite the use of drawing rather than written language, their levels of literacy made the 'paper and pen' approach to mapping difficult. The women became the map. Significantly, these women retold this experience with a pronounced sense of ownership, describing the activity as one of the participatory methods that they themselves had created.

This experience illustrates an example of PAR that facilitates a way of learning and communicating knowledge conceived by all participants, as opposed to research participants' involvement being pre-conceived and pre-determined (see also Sanderson and Kindon 2004). In relation to mapping, this means participants are not simply mapping out a pre-determined cartographic frame of reference: they inform the cartographic frame. In relation to the example above, the cartographic frame finally created privileged embodiment rather than two-dimensional abstraction.

The underlying significance of approaching cartography within the philosophy of PAR is seen in the prominence of cartographic language within contemporary social theory. Spatial language and spatial thinking is strongly emphasised within many emergent strands of philosophy (see Crang and Thrift 2000). Specifically the language of cartography is attributed to particular philosophers to emphasise that such philosophies conceptualise, or seek to conceptualise, the world in a new way (see, for example, Lorraine 1999). These uses of cartography indicate towards its ontological rather than methodological use – its ability to determine how we see the world. Iimplicitly, PAR has the same desire. That is, to create collaboratively knowledge that reflects the knowledge and communication of all those participating and consequently, as a function of that participatory process, to see and communicate the world in a new way.

Participatory cartographies therefore approach research participants as latent cartographers and also recognise the cartographic role of the researcher. Because the reference to cartography rather than mapping is an emphasis on the ontological level of participation sought, recognition must also be given to the epistemological and methodological complexity of this knowledge production within PAR, which is now well documented (for example, Cornwall 2006a; Lennie 1999; Mosse 2000; Parfitt 2004). For example, not all cultures or genders have the same ability to inform dominant means of conceiving and articulating knowledge (see the critiques by Bhabha 1994; Irigaray 1985; Spivak 1988). The epistemology of participatory cartographies is distinguished from participatory mapping by an intentional concentration upon the ontological perceptions informing cartographic frames of reference.

Participatory cartographies can therefore work with existing PAR mapping methods, but distinctly seek to make explicit and negotiable the ontological assumptions within those methods prior to their potential use. As a consequence participatory cartographies may produce two-dimensional 'maps' or diagrams, but not necessarily. They will take various forms, reflecting the manner of perception and communication prominent among participants. A performative map, such as the example cited above, is created when participants intentionally perform and embody the relationship between people, place and space that the research seeks to articulate.

Figure 15.1 represents a conceptual outline of participatory cartographies. This suggested framework is for researchers or practitioners seeking to engage with participatory research partners around a research topic that contains either an explicit engagement with space and place relations or seeks to develop more collaborative mapping methods. The outline integrates different qualitative research methodologies with the intentional aim of methodological innovation by and with research participants.

| *Methodological approach* | *Cartographic emphasis* | *Theoretical significance* | *Distinct from cartography 'as usual' which is...* |
|---|---|---|---|
| Qualitative interviews and narrative enquiry | Storytelling around space and place, specified to a particular research interest, yet simultaneously dialoguing the conceived connection between particular people and particular places | Seeking to articulate experienced but subjugated cartographies within the space or place the research is focussed upon | Based upon an external cartographer's pre-determined cartographic assumptions |
| PAR | Collaborative design of the cartographic frame of reference: determining methodologies, techniques, appropriate language/words/symbols to use in conversations and discussions; determining who should participate; participant analysis of results | Seeking to explore understandings of space and place in participants' own terms and visual representations or symbols | External determination, and then extraction of, cartographic data |
| Research as performance | Exploring cartographic representations through a variety of creative means – role play, theatre, song, dance, drawing and diagramming – with subsequent discussion around these performances | Exploring a range of 'languages' through which participants can communicate their cartographies individually and collectively | A two or three dimensional map corresponding to the technical language of the cartographer |

*Figure 15.1* A conceptual outline of participatory cartographies (Source: author 2006)

To summarise, participatory cartographies explicitly engage with the challenges facing PAR mapping practices. Namely, they take seriously the recognition that mapping is not a value neutral activity and consequentially ask how we can work with participants to map in a participatory way that does not determine *a priori* what mapping constitutes methodologically.

I now turn to reflect on my own participatory cartographic experiences within two community groups and highlight some challenges and potentials for this approach.

## Cartographic performances with communities within Fiji and Tanzania

The context from which the following reflections are drawn is a doctoral research project exploring cartographies of development space, with a specific focus on the experience of embodying development and spirituality. It involved two place-based community groups with an explicit Christian spirituality and engaged in development work. The first group were members of a Melanesian settlement and Anglican parish in Fiji, which I refer to as Holy Family Settlement (HFS). The second were members of Mothers' Union groups in an Anglican diocese in Tanzania, which I refer to as Umaki. In this research context, the cartographic framework was important to enable open articulations of perceived spiritual spaces informing community members' engagement with their placed-based development work as well as openness to socio-cultural perceptions of space and place. Our cartographies were not created for a specific external audience but primarily were part of exploration and communication amongst ourselves.

The first challenge that needs to be highlighted with this approach is that the language of the framework above, or the specific academic language of the research project, was not necessarily the language of communication within the participatory interactions. Finding a shared language for the research is an important part of relationship building with participatory partners and creating cartographic frames of reference. Within HFS we spoke of creating a map of the settlement – of the settlement now and in the future – through our shared stories and visions of 'this place and this future place'. Within Umaki, the placed-based emphasis was changed to an emphasis on the actual Umaki members, their 'herstories', their relationships within the groups and their prayers for Umaki's future. The research focus did not change between these two participating groups, but the language and frame of reference for the research did, which provided valuable research insights.

### *Holy Family Settlement: weaving narratives, writing places*

The gathering of stories, and the way of mapping initially chosen, was lead principally by a self-titled research team of three women within the settlement. In addition the process was guided by the male priest and male chief of the settlement. In both cases, our gathering and our commissioning occurred around weavings from the community. Ceremonies of welcome and our conversations took place upon woven mats. The way of representing our maps of HFS became based on these weavings and we used the ceremonial mat as a pictorial depiction of the settlement in a series of posters. Each poster/map depicted conversations upon a mat, as they were in practice, with each thread representing a different thread of thought or opinion. In this way, both the diversity and unity perceived within HFS were expressed.

The community aspect of the weaving also captured a strong sense of place: HFS being a place constructed and created from the people of the place. As our participatory cartographies also reflected spiritual and communal activities, the woven depiction of our 'maps' was particularly appropriate given that within the Pacific weaving is understood as a community and spiritual activity (Johnson and Filemoni-Tafaeono 2003). Within HFS, the most explicit discussions around the cartographic framing of development space took place in conversations around a diagrammatic representation of the three most significant threads within our conversations and their interconnection. These threads were: material and physical needs, community and culture (represented as *koro*, meaning Fijian village), and spirituality. Participants stressed that development within HFS required these three aspects to be connected.

The separating out of 'material needs', 'community and culture' and 'spirituality', which I had initially done to draw out an overarching understanding, were false delineations as the female participant's statement in Box 15.1 illustrates. This is because these three aspects were considered to be based in the person, as was HFS, and therefore unable to be separated. So the cartographies of development space for HFS were also the embodied cartographies of development space within the people of HFS. These understandings of the connections between people and places within this particular context have important implications for approaching these research participants as cartographers.

Maps create connections between people and places, as already expressed. Participatory cartographic performances represent and create those connections. Within HFS, the series of posters, (our maps), included 'Our Future Vision' and 'Realising our Future Vision'. These particular maps represented the envisaged future place of HFS, and play an instrumental role in creating that future place. Through our focus groups around the posters, it was decided that every household in the settlement should have a copy of the posters in Fijian (Box 15.2).

**Box 15.1 Connections**

The relationship between the spirit and the community/*koro*: we are all individuals but one in God. We have different roles, but we are all one in God. So all these three things are not separate but together – and the *koro* is from the people of Holy Family Settlement.

Source: Female HFS research member

**Box 15.2 The poster**

I will put this on my door, so that every morning when I go out I can see that this is our vision.

Source: Male HFS research member

Crudely put, maps can show how to get somewhere, and our collaborative mapping maintained something of that characteristic. The maps were a prompt for activities within the settlement, such as the building of a community centre and the establishment of regular settlement meetings; they were creative as well as representative of the connections between the people of HFS and the place HFS.

Our means of collaborative cartography was met with an enthusiasm understood, in part, by appreciating the settlement's wider political history. The uniqueness of the Melanesian settlements in Fiji, which were formed following the period of indentured labour of Melanesians to the sugar plantations, was highlighted by the lack of a Fijian word to describe this unique settlement (Box 15.3).

The enthusiasm expressed around the posters and the gathering of stories that formed them, came from participants expressing unique connections between particular people and a particular place, connections which 'cartography as usual' could easily miss (Plate 15.1).

### *Umaki: weaving together hearts and homes*

Weaving was again an activity prominent in the collaborative cartographic performance in Tanzania, albeit differently from within Fiji. Weaving was an income generating activity amongst Umaki and I was given a woven basket as my leaving gift from them. Weaving was a cartographic frame of reference for Umaki, expressed in the way that they wove together their communities and the way I and my companion Ruth Newport[2] were also woven into their community. Their chosen method of 'teaching' me about their development, was for me to become incorporated into their Umaki activities.

**Box 15.3 The limits of words**

One thing here is the word … settlement and the other is *koro*. Holy Family Settlement is a settlement, different kinds of people from different *mataqalis* [tribes] that settled together and have come from different places. *Koro* and the village, they are only one kind of people and they have got a lot of land. Here we've got no land. For us, we can use *koro* because for us we don't know how we can translate settlement into our language.

Source: Male HFS research member

*Plate 15.1* The Holy Family Settlement research team working with the posters (Credit: Eleanor Sanderson 2005)

One aspect of the research 'plan' with Umaki members was to create a *cheza* (meaning dance, song or drama) as a way of expressing their development space. They chose to do this by enacting a Bible story, which they concluded with the self-expression articulated in Box 15.4.

This expression of the heart and home became the recurring theme around which we learnt from Umaki (see Plate 15.2). This was a teaching about being part of their homes and being part of our hearts. In leaving Umaki the woven basket I was given as a gift was expressly intended to inform the practice of my home and the content of my heart. Our cartographic performances, therefore, were to be inscribed within me as I became the one who would teach others about Umaki and their cartographies of development space.

**Box 15.4 Expressing development space**

Our hearts are our homes and our homes are our hearts. If our homes are not open to the Spirit of God then neither are our hearts and if our hearts are not open to the Spirit of God then neither are our homes.

Source: Umaki research participants

*Plate 15.2* Members of Umaki, 'My Heart' written on the wall of their house (Credit: Eleanor Sanderson 2005)

Our means of combined analysis was a participatory ranking exercise using pictures that brought together the different aspects of Umaki's development space and prompted discussion about their relative importance and their interconnection. These aspects included craft projects, Bible study, income generating activities, prayer, literacy learning, *cheza* and visiting. The linear framework of the ranking was repeatedly subverted by Umaki members. In one case the women placed all the pictures that showed their easily identifiable development activities on top of those showing more spiritual pursuits. In another case, women drew two different pictures of 'prayer' and placed them at either end of the ranked pictures to indicate that prayer surrounds all aspects of their work, and informs all aspects of their development space.

In addition to this subversion of standard participatory techniques, each of the participatory ranking exercises was concluded by praying for the future of each of the aspects of Umaki. These prayers were part of the cartographic expression of the group. The prayers articulate and perform connections between the particular people and places of Umaki and the way that these connections are conceived. For example, the places woven included the community from which each Umaki group takes their name, the houses and hearts of each Umaki member and the place articulated as within 'the hands of God'.

As with HFS, the collaborative and intentional desire to envision and express particular perceptions of our world can provide both fresh insights into areas of research and also unsettle recognisable frames of cartographic reference. For example, the language of cartography at times seemed inappropriate to use in the context of our research within Umaki because these women had low levels of literacy and were unfamiliar with maps, and because our research often did not seem recognisably cartographic – where were the maps? Like those in Fiji, they were performative. Our research practice became an intentional cartographic performance directed by participants. The connections between people and places and meanings of the spaces of development were mapped out through our performative research activities as space was created for participants to show us how. This process was (understandably) often confusing and placed significant intuitive demands upon us as researchers (Box 15.5).

Despite this sensation of exhaustion and confusion, what emerged through our performances was a clear sense of ownership by the women and an explicit expression from them that we 'came to them and met with them', which enabled them in turn to be 'drawn to us'. Through this process we therefore developed an understanding of the cartographic frame of reference within which development was negotiated by these women, principally because this was the frame of reference within which our own interactions were negotiated.

## Where to next? Encouragements for future participatory cartographers

These two case studies, although sharing the same research focus, illustrate the different forms that participatory cartographies can take: one illustrating two-dimensional diagramming, the other explicitly performative cartography. These distinctions indicate the transferability of this methodology into very different community groupings and potentially different research settings. These cartographies were focused around our own exploration and communication and were not created explicitly for external use. In so doing, the forms these cartographies took did not necessarily capture the potential power of already accepted mapping practices; therefore they might more easily be ignored or dismissed by people of influence outside these communities. Yet in this research

### Box 15.5 The 'blah' of embodied methodology

Blah: I think there is no other word that best describes the methodology better than 'blah'. The uncertainty inherent in the research approach and the highs and lows of the whole process also contributed to the 'blah'. The confusing mess of stories that at times seemed contradictory often resulted in a feeling of 'blah'. The inherent involvement of the entire body of who we are and the relational nature of the approach means there is no tidy end or closure – no precise definition to the process, it is just 'blah'!

Source: Ruth Newport, written reflections

**Box 15.6 Changing ourselves**

You have bound me into your way of creativity. Without explicit speech, you enfolded me into your way of expression, your way of gathering together. I weave with you, following your rhythms and gradually releasing the anxiety caused by my trained need to disentangle, separate out and name the individual parts.

Source: Eleanor Sanderson, written reflections

context they facilitated the ontological shifts necessary for me to more effectively engage with the relevant wider socio-political environments of concern to myself and these communities. Nevertheless, I believe it is also feasible and desirable for participatory cartographies to explicitly seek to collaboratively create cartographies for external uses as part of accessing the powerful dynamics of mapping practices.

The democratisation ideal within PAR inevitably challenges normative assumptions surrounding research processes (see also Kindon *et al.*, Chapter 2 in this volume) and participatory cartographies invite such challenges around the conception and methodology of mapping. These challenges are not always easy to experience as researchers because they are de-stabilising of our supposed knowledge, authority and power. Engaging in PAR acknowledges the need for different people's perspectives to shape research that reflects and informs their lives and challenges us, as researchers, to facilitate that process. This can and should result in changes to our own perspectives (Box 15.6).

## Notes

1 This chapter reflects the collaborative creativity of these individuals named anonymously or directly. Whilst written by Eleanor Sanderson, it incorporates reflections by other collaborators.

2 Ruth is a post-graduate researcher in *Aotearoa*/New Zealand who was born in this area of Tanzania and lived amongst Umaki for twelve years during her childhood.

# 16 Participatory art

## Capturing spatial vocabularies in a collaborative visual methodology with Melanie Carvalho and South Asian women in London, UK

*Divya P. Tolia-Kelly*

### Introduction

Participatory Art as an approach to research situates the 'visual' as 'inextricably interwoven with our personal identities, narratives, lifestyles, cultures and societies, as well as with definitions of history, space and truth' (Pink 2007: 22) and thus can be an appropriate, effective tool of qualitative research. The engagement with visual methods can also be the means to work across disciplinary divides for academic research (see also Chapters 14–20 in this volume). Visual processes can be used to 'triangulate' qualitative research methods, developing a richer relationship with views, politics and experiences beyond the restraint of written and oral practices. Visual methods have long been used by those wanting to engage with the 'experiences' of those marginalised within society, for example children, women and people with mental illness, and by researchers of rural development policy in the majority world. The politics of using Participatory Art enable the research to be engaged ethically, plurally, creatively and inclusively towards the development of theory, policy and practice.

The use of Participatory Art has two critical aims; firstly as researchers we engage with textual communication that offers 'voices' and 'perspectives of participants in a visual process' (see Box 16.1). Secondly, the process of engaging with 'visual' communication goes beyond that which written questionnaires and oral interviews can engage, as participants physically produce maps, drawings, technical designs, photographs or videos themselves (see also Chapters 14–20 in this volume). Making voices and perspectives tangible in a visual form adds scope for unexpected or new grammars (constellations of words and meanings not usually encountered or expected by the researcher given their different social positioning, views or background) and vocabularies that are sometimes inexpressible in other contexts.

In multilingual research, Participatory Art can bridge the 'gaps' in linguistic understanding. For example, even individual words such as 'landscape' are problematic and hold complex plural meanings amongst English speakers; involving American-English and Indian-English speakers in research on 'landscape' may compound differences further. What a visual process can do is to make tangible and record experiences, meanings and views in alternative forms. This process creates visual texts that are a means to 'ground' theory, policy and practice, through a process that allows space for an alternative form of expressing experience that builds bridges across difference.

With the collaboration of an artist, the dissemination of research can also be extended beyond written text in the form of exhibitions and photo-essays (see also Krieg and Roberts, Chapter 18 in this volume). Although the artist is not always essential, in this research I felt that it was important to work with a professional to achieve an aim of the research, which was to produce art for exhibition in a formal gallery space. The collaboration with a professional artist enabled rigour in

## Box 16.1 Participatory art

**Aims**

- To exchange ideas, values and experiences through visual communications (participants may produce art within various 'genres' of the visual, including mind-maps, drawing and design)
- To create a space for embodied, multilingual, marginalised experiences to be expressed through visual form
- To produce a set of visual materials for analysis leading to grounded theory, new understandings of experiences, vocabularies and views for policy development, participatory action, and/or social knowledge

**Process**

- Creating an empowering participatory research space and process where trust and communication is established - where liminality is achievable
- A group process where the participants are involved in recording their views and experiences through a visual process of making, analysing or producing collages of drawings, pictures, maps or photographs
- Working with an artist in a collaboration that challenges and/or extends disciplinary understandings
- Working in groups using 'ready made' images without an artist practitioner

**Outcomes**

- Recording experiences beyond written and verbal texts within a participatory process
- Dissemination of research in visual texts and exhibitions
- Developing 'grounded theory' based on extending 'the space of exchange' available to both researcher and participants
- Opening up the terms of engagement with social groups beyond preconditioned social expectations of the researcher, leading to better understanding and development of new vocabularies and grammars for research, communication, policy and practice

Source: Author's own experiences

producing visual materials that are socially and culturally recognised as 'Art', and provided essential advice and skills necessary to avoid the risk of participants viewing the research process as naive, 'experimental', unethical or patronising.

If action learning is to engage with people in a grounded way, then visual vocabularies recorded through a participatory process can broaden the terms of engagement and also act as a communicative and educative tool for both the researcher and participants (see also Stuttaford and Coe, Chapter 22 in this volume). The tensions of *translation* and *mediation,* however, remain. The texts produced through Participatory Art require visual analysis and remain texts from which meaning is garnered (the researcher is continually mediating the meaning of a visual text), in the same as in any other qualitative research process. In the collaboration between artist, researcher and participants, meanings, experiences and views are constantly being negotiated and mediated through verbal, written and visual forms.

In my own research I have investigated postcolonial 'identity' amongst migrants living in the United Kingdom (UK), in particular those who are politically 'British' and culturally 'British Asian'. My aim in using visual methods is to carve out a research space, and ultimately a dialogic space (where learning and communication is two-way), which allows room for migrants to discuss the experience and value of 'landscape' in relation to their sense of belonging and identity. Landscape has been critical in European and North American cultures for securing 'cultural identities' (Schama 1995); however, these cultures do not reflect the values of all communities, including for example, working-class perspectives and those of children and young people. Participatory approaches essentially encourage a *liminality* (a result of the experience of a research process that is participant-centred to the degree that the process itself is unobtrusive) which allows for participants to feel relaxed and 'free' to express themselves, to shape and challenge particular assumptions within the research process, and to introduce terms and vocabularies that are more relevant to their experiences. To enable liminality, the researcher's own *reflexivity* has to be constant within the research process, to enable the participants to become empowered enough to take ownership of the sessions and the terms of debate itself. Reflexivity is about the researcher continually assessing power relationships within this process, and being self-critical about his/her assumptions, judgements, expectations and self-perceptions.

## Participatory process in the 'Describe a Landscape …' project

### *Politics, concepts and theory*

In this section I want to work through my own experiences of research using a participatory approach with landscape artist Melanie Cavalho[1] and South Asian women living in London. In our work, we used a visual methodology that was also a continuation of the project by the artist, entitled 'Describe a landscape …'. The sessions with the Asian women were designed with the artist.[2]

The decision to use art and an artist with participants was for me a political one; to produce a tangible record of participants' landscape values not normally encountered in art exhibitions. The participatory approach countered the fact that 'landscape art' is usually embedded in a network of elite producers of paintings, sketches, drawings, diaries and poetry. Many of these texts fail to engage with everyday perceptions and values of working-class, ethnic or other marginalised communities.

In working with the artist, what I aimed to do was to make tangible on canvas a different set of vocabularies and visual grammars that operate beyond this elite lens. I sought to include what is often 'not looked for' or what is edited out of the western picture of 'landscape culture', and aimed to record landscape relationships formed outside of the European context. The postcolonial relationships represented in this process reflected the hybrid quality of the participants' citizenship. The research also addressed the lack of academic inquiry into the 'circulation' of landscape values throughout the colonial geographical footprints of European governance.

### *Working landscape and art: collaboration with Melanie Carvalho*

In the research design we decided to recruit and run small-group sessions with South Asian women living in London, all of whom were first generation migrants. We recruited ready-formed, mixed groups of women from a variety of ages and backgrounds from two Asian centres in West London. The group comprised women who spoke Urdu, Punjabi, Hindi, Gujarati and English. The multilingual approach in using each woman's first language seemed to be the most ethical way of working with these participants that would ensure their comfort and reduce obstacles in talking through abstract concepts such as 'landscape' (see Gluck and Patai 1991). This approach enabled the group to 'own' the sessions; an empowering process suitable for working with marginalised and minority communities (see Burgess 1996; Burgess *et al.* 1988a; 1988b; Holbrook and Jackson 1996; Kitzinger and Barbour 1999).

Holliday (2000) argues that innovative visual methodologies can counter the traditional power dynamics of other methods, but require continued reflection on whether their aim is to 'further legitimate the truth of the research itself' (2000: 504). Pink (2001), in response, argues for reflexive modes of representing participant 'voices'. In this research and its dissemination, the images produced were always situated within a biographical context and the process of production outlined as transparently as possible.

The move towards working in a multilingual space beyond English opened up the possibility of forging more even relationships within the group, and making the 'doing' of art and painting possible. The use of art materials allowed us to attempt to capture alternative vocabularies and visual grammars that are not always encountered or expressible in oral interviews. In previous research it has been difficult to get usually more conservative South Asian groups to talk about abstract environmental values. The process of abstracting 'environmental values', 'emotional values' and 'aesthetic responses' was assisted by being with an artist in a group situation enrolled in the unusual physical actions of using hands, fingers and arms differently to the day-to-day (Bingley 2003). It was difficult and cumbersome, as encouraging people to paint, draw or talk about landscape and emotions inevitably is, but we tried to counter problems by creating an environment where the women 'owned' the space. All the sessions were recorded; together these visual and aural texts formed the basis to Melanie's paintings.

The images produced were in turn used as sketches by Melanie and were the resource for producing a set of paintings, which formed a record of the women's landscape values and experiences (Box 16.2). The aim of working with an artist was to elevate the project to being about creating an alternative archive in the form of a set of paintings, which were publicly exhibited at the University of Plymouth (Royal Geographical Society Annual Conference) and in London (Cubitt, and University College London).

### *The women's reactions to the exhibition*

The exhibition in July 2000 at University College London allowed the women themselves to view the paintings in a gallery situation. This private viewing of the final pieces allowed the women to continue the process of empowerment. Their comments and reactions were recorded and each participant also received an A4 print of their final painting. This was a nervous moment for both the artist and myself; the women had entrusted us with their intimate descriptions of real homes, personal fantasies and some projections that combined the two. For Melanie, it was the first time that she had received verbal feedback from the people who had contributed their descriptions to her.

### Box 16.2 Extract from interview with Melanie Carvalho

MELANIE CARVALHO: I try to do the paintings in one sitting. The idea is to be as objective as possible and respond as technically as possible, without elaboration. To create a romantic landscape painting from someone else's imagination is really exciting. As you'd expect the way people responded was really varied, a mixture of writing and drawing. Some people responded very abstractly. [...] For me that's what was empowering about this project. By making the paintings, they then have a record of their history.

PAUL ANDERSON: What do you make of the results?

MELANIE CARVALHO: The difference between these paintings and the ones I've done with other people is quite dramatic. A lot of women described their childhood home, a place where they'd actually been. Many of them worried that their drawings would look childish, and wanted me to make them more picturesque. They didn't look childish to me; they looked beautiful.

Source: Anderson *et al.* 2000: 112–19

Overall, the women were very impressed. Shilpa described her painting as a really beautiful image and said that it reminded her of the desert in Sudan where she had lived. For Shilpa, it was a true representation that reminded her of her life there (Plates 16.1 and 16.2).

When another woman, Puja, saw her painting she said, 'This is the place where I played as a child'. For her, the canvas was not a metaphor but a representation that *was* the place. She was tearful at the sight of it; she had not been there for over a decade. The emotional response of many of the women was captured in Kanta's comment, 'For me, you know, this really gets at, I suppose in a way, what I feel inside' (interview at exhibition, July 2000).

Lalita, another participant, was also very pleased with Melanie's representation of Nainital, as for her it represented a very precious time in India (Plate 16.3). The combination of water and greenness along with peaceful surroundings was a recipe for security and serenity, which Lalita was trying to recapture wherever she went.

Melanie and Lalita discussed the ways in which specificities of place could be treasured, but also rendered unimportant when compared to the recapturing of sensory memory that new landscapes can evoke (Box 16.3).

For Lalita, my own reading is that her canvas provided a zone that was spatially and temporally in flux. It resonated with her own dual connectedness with both the UK and India. The lake was an iconographic memory, at the sight of which she recalled two places at once. The symbol is iconographical because in both Derwentwater and Nainital there were other features that she had not described. For example, she said that in Nainital there was a development of houses along the whole perimeter of the lake.

*Plate 16.1* Untitled landscape (Shilpa) oil on canvas (original painting produced by Melanie Carvalho 2001)

*Plate 16.2* Shilpa's description (Credit: Shilpa, workshop participant 2001)

*Plate 16.3* Untitled landscape (Lalita) oil on canvas (original painting produced by Melanie Carvalho 2001)

**Box 16.3 Exchange between Lalita and Melanie Carvalho**

LALITA: I have noticed that with landscapes, they jump down on your mind from childhood. Wherever in life that you move, you try and locate if possible a place to co-relate or to you...Know it's just there... You look for it wherever you go in the world. I just love it. I just like it. When I came to England it was the first thing I want, I want to be near a lake.

MELANIE: So I gave it a sort of Northern light, because so many people had a tropical light... So that you couldn't really tell whether it was India or...

LALITA: A landscape can be from anywhere, anyway.

Source: discussion at exhibition July 2000

## Conclusion

It is important to recognise that the kind of research described here serves researchers' agendas. However, researchers can aim for empowering, participation-centred research and a reflexive approach throughout. The participants involved here were not mined to produce images, but were enabled to share their 'positioning' in the world through the practice of Participatory Art. Admittedly, the project did not effect deeper structural change, for example in how 'British Art' or 'Englishness' were defined. However, the process itself was ultimately empowering of the 'voices' of those British citizens not normally included in formal art spaces and whose perspectives are certainly not reflected in academic literature on 'landscape'. The resulting exhibition made material *their* perspectives and voices. Thus the 'progressive' gains of this project can be measured via the political synergies enabled between the collaborators, and in the chemistry between them, and the building of trust, credibility and respectful engagement; being a reflexive researcher was just one cornerstone of this participatory success.

In visualising and revisiting the cultural narratives of transnational communities that are living within Britain as a result of postcolonial mobility, their landscapes of memory and/or history have been sought and made tangible in the 'Describe a landscape…' project. The paintings produced illustrate how landscape is a relevant and conceptual tool in the everyday lives (see Anderson *et al.* 2000) of migrant communities in London. They also re-inscribed the cultural record with their voices and perspectives. This way of working is in the vein of writers like Berger (1972); Jazeel (2005); Kinsman (1995); Mitchell (2002); Nash (1994; 1996; 1997) and Rose (1997; 2001), who counter the traditional elitist and sometimes masculinist orientation of landscape research by using collaborative methodologies to shift the registers of the debate.

## Notes

1 Melanie Carvalho has been trained at St. Martin's School of Art, the Royal College of Art and held a scholarship at the British School at Rome. She has curated and exhibited in several international exhibitions including at the *James Brown Galleries* (New York), *Cubbitt* (London), *Hidde Van Seggelen* (London) and has published a sole-authored book entitled *Expedition* (2006).

2 What became clear through this process is that artists have their own varying relationships within the art world. The issue of whether they collaborate or not with researchers rests on their investment, usually balanced between their commercial interests in the network of galleries collecting and selling art, their need to be evaluated within the regimes of art theory (based on being in art collections, represented by galleries, and being in particular art shows) and their commitment to the intellectual basis of the research process. In the collaboration with Melanie Carvalho, we had similar research interests in landscape and memory and this drove the project forward.

# 17 Participatory theatre

## 'Creating a source for staging an example' in the USA

*Marie Cieri and Robbie McCauley*

### Introduction

For both of us (Robbie, an African-American theatre artist and academic, and Marie, a white American arts producer turned geographer and academic), the creation and presentation of theatre is a site of Participatory Action Research (PAR). Artists and researchers in collaboration with diverse individuals and groups can produce Participatory Theatre (PT) to explore questions of importance to their communities and to initiate a process of dialogue and interaction that can effect change at a broader scale. In our experience, the means and aesthetic of PT can play a vital role in changing how people think about, and act upon, social and political relationships within their communities.

Our work in PT is typified by the *Primary Sources* series, which took place in three areas of the United States in the 1990s and emanated from a shared impulse to talk about race through theatre. *Primary Sources* took several large, charged and remembered events (the early 1960s voting rights struggle in Mississippi; the mid-1970s school desegregation controversy in Boston; and the 1969 Black Panther Party–Los Angeles Police Department conflict in Los Angeles) and sought to explore the many important but untold stories surrounding them.[1] Collection and discussion of these stories formed the basis of our work in collaboration with residents of various racial and/or ethnic, class, age and educational backgrounds in each location. In previous work, as well as for *Primary Sources*, we developed repeatable processes and guiding principles informed by sustained artistic practice and shared interest in the subject matter (see Boxes 17.1 and 17.2).

Much of what follows, however, is excerpted from our dialogues about PT in Boston in July 2006, mirroring the fact that dialogue forms the basis of our work together as well as with others.

### In the beginning

ROBBIE: It was the late 1980s, and I remember being stunned when you said race is something white people don't talk about. I knew, and didn't know, that was true. My prejudice made me think, 'It's something white people *dare* not talk about.' White people, I thought, were still walking around not having dealt

**Box 17.1 Typical participatory theatre processes**

**Origins**

- PT is grounded in sustained artistic practice
- PT artist-researchers are dedicated to making theatre informed by material compelling to participants and their community or communities.

**Preparation**

- Identify research location and topic
- Talk to local theatre people and potential community partners
- Seek funding and in-kind support (actor-collaborators will be paid for their work, affirming the value of their contribution and the community stories they tell)
- Basic tool: audio recorder

**Script development**

- Recruit actor-collaborators through people you know or locate at local arts, educational and community organisations
- Start PT with repeatable processes:

  'Talk-abouts' focus on the event and why it was chosen

  Actor-collaborators and artist/researcher(s) record interviews with local citizens who have stories about the event and its aftermath

  Hold story circles where collaborators dialogue about the stories collected and add their own material (minimum 12 hours per week for three weeks), all recorded.

  Scripts are developed from transcriptions and shaped by artist/researcher(s) in collaboration with actors in a narrative collage informed by jazz rhythms.

**Rehearsal, performance and purpose**

- Rehearsals proceed from talk-abouts (minimum 12 hours per week for two weeks). Music, movement, set design added as budget allows
- Actor-collaborators variously play themselves as well as other characters
- Modes for addressing the audience with subtlety and/or directness are proposed. Goal is to engage audiences theatrically as well as inspire more stories and dialogue about the targeted event and its legacy
- Performance spaces are secured in places where stories resonate
- With the help of actor-collaborators and others in the community, word is spread on the ground and in the media about the performances

Source: Authors' own experiences

with their internalised sense of supremacy. But I found myself speaking my thoughts out loud, and you were listening intently. I was vulnerable to the questions of class and individuality you were raising. We were two mature black and white people having an intense public conversation about race in a restaurant. Theatrically for me, the simplicity of what we were doing – the conversational music we were playing – was inspiring.

MARIE: I was at a conference recently where a woman spoke about wanting to create bridges between Muslim and white working-class youth, but not knowing how to approach those two groups as a white academic. I've found it's important to have a partner to connect me to the issues in a way I can't by myself, especially if I'm trying to cross lines of social, political and economic difference and truly participate in a movement towards positive social change. Her dilemma made me think about how you and I started working together. There had been things about racism in my life I had wanted to address, and then I saw one of your performances, and it was both so political and so personal. I knew then that I wanted to work with you if you were willing. And

## Box 17.2 Principles that guide us

These are not hard and fast rules, but ones that can inform PAR as practised through theatre. Such repeatable guidelines are needed for the process to do what it will and to release us from focusing on the end product so much.

**Commitment and respect**

The artist-researchers and the community commit to the process and agree to consider differences and transformation as core values.

**Humility and connection**

Stories are celebrated as being both personal and bigger than any one individual (i.e., it's not just your story).

**Subjectivity and objectivity**

Subjective understandings and perspectives are as valuable as apparent objectivity.

**Listening and hearing**

These strenuous and essential practices heighten the physical, mental and sensory aspects of being present with, and paying attention to, what others are saying. Attention to content and presence with one another make for dynamic dialogue that can crack walls between people. Wilful acts of listening and hearing are transmissible to audiences and can provide an example that can transform behaviour.

**Collaboration and exchange**

Personal exchanges about charged issues during the making of the work are welcome and must be grounded in respect for differences. Such exchanges provide examples for actor-collaborators, researchers and audiences to continue the conversation beyond the performance space, thereby enhancing possibilities for creative social change.

Source: Authors' own experiences. For more guidance on PT, see Boal 1992; Cieri forthcoming; Clark 2002; McCauley 1996; and Rohd 1998

I think the partnership we developed, without us expressly calling attention to it, became a model for our collaborators. Our ongoing dialogue might have made it easier for them to enter into, and engage with, the issues in parallel ways.

ROBBIE: I think PT has to involve each person's own encounter with the subject matter. Our expertise came not only from our past work as artists and our interest in charged material, but also from participating with everyone else in the process of researching and making theatre (Plate 17.1). You, as the producer, were part of those story circles in Mississippi, and whoever was with us – eventually the light people, the sound people – got connected to the subject matter, and some of their stories found their way into the script.

## Dialogue

ROBBIE: Dialogue [like in Box 17.3] helps the actors connect with and know each other. It allows the material collected through interviews and story circles to be shared. Sometimes the dialogue that shows up in the script is a result of placing material together aesthetically, jazz-like. But the staging of dialogue among the actors is an attempt to provide an example for talking about

*Plate 17.1* Some Mississippi Freedom actor-collaborators and story souces on a road near Jackson, Mississippi, 1992. Left to right: James Green; Veronica Cooper; Leroy Divinity (story source); Kay King Valentime; Deborah Imboden; Sameerah Muhammad (partially hidden); Kent Lambert; Clarie Collins Harvey (story source); Sadat Muhammad; Sheila Richardson (Credit: Marie Cieri 1992)

**Box 17.3 Sample from the script of *TURF: A Conversational Concert in Black and White***

(Actor-collaborators Paula and John were African-American and Irish-American respectively, and Robbie appeared with them in this scene from the Boston segment of *Primary Sources*.)

PAULA: I'm not talking about it as a moral thing or a we-just-got-to-put-up-with-it thing. I'm talking about it as something that is possible. I mean, people sit down and say, 'the world looks like this, why can't my neighbourhood look like this?'

JOHN: Now I can understand other viewpoints. And I would have no objection to seeing changes. But like you say, there is that fear of having such a radical change that it may have such an effect on an individual, a community. As a white person, you'll feel threatened perhaps if you felt uh... like a lot of people fear that the whole South Boston's gonna be black soon. Roxbury at one time... used to be a white community. And I guess the thing is, uh, everything black people touch... goes bad.

ROBBIE (WITH IRONY): Thank you.

Source: Robbie McCauley

charged social, political and historical material that is often not present in regular discourse. We're creating a source for staging an example.

MARIE: Sometimes there's contestation among the actors, between you and the actors, between actors and possible interviewees, and between potential performance co-presenters and me. But this is not something we want to erase.

ROBBIE: We want the dynamic of dialogue around charged issues. However, we're not looking for winners and losers. We're looking for the textures, the rhythms and the dynamics that the material ignites for theatre. We want the audience to be open to what is said through the beauty of the dynamics.

MARIE: But what happens when some of the actor-collaborators don't want to say something on stage because they think they'll appear politically incorrect, or just wrong? How do we bring that disagreement into the participatory mode of creating the work?

ROBBIE: This mostly came up in Los Angeles, and the challenges were around questions like: Are you 'telling it right'? Who's got the 'right, authentic story'? We work at the tensions around authenticity and appropriation as we develop the performance.

MARIE: But there's no final answer given on stage.

ROBBIE: A colleague once said to me that authenticity is everything and authenticity is nothing. It's really about the engagement, the telling. That's why, in this process, the actor-collaborators have to deal with their own questions about who they are to tell the story. They have to come to terms with the fact that they are not just actors telling the story; they are who they are, with the actor's craft, telling the story.

MARIE: The point is not to get up and tell a single 'right' story. The point is to put often difficult or challenging aspects of community dialogue on stage, to engage audiences in that dialogue and for it to continue after people leave the theatre through the example of what's been put on stage. Then there's the statement often made to us that 'It's better not to talk about it'. In Los Angeles, especially, some felt that talking about law enforcement and the Black Panthers would be incendiary, and some people didn't want *TURF* to be performed in South Boston because they thought it would result in renewed racial violence there. But remember what the director of the South Boston Boys and Girls Club said? He said the young people who come to the club are living in a community whose population is changing, and they're going to have to deal with that sooner or later anyway, so why not through theatre?

## Talkbacks

MARIE: Let's discuss the post-performance talkbacks, how they are part of the PAR, how they circle around to affect the presentation on stage, how they're also part of the stories being told.

ROBBIE: We create ways for the actor-collaborators to be present within the talkbacks that are more than just being there. In fact, I'd like to increase the performance aspect of the talkbacks, so that the actors, once they've finished a shaped script, step into a time of improvisation. Then, as in a sing-along or jazz concert, the audience might be engaged in a way of listening and responding, of saying things that are inspired. That happened in Port Gibson, Mississippi, when a minister in the audience got up and quoted scripture in a way that was particular to *Mississippi Freedom* and to which the actors almost said 'Amen'. So what we're looking for is how the aesthetic of PT allows a community to initiate the retelling of its stories, particularly those associated with pain and grief, struggle and triumph. We haven't measured that. But I would bet that much has resonated from those moments of retelling, because unlike television there's breath in them.

MARIE: Are you ever disappointed that the performances don't have a longer run, don't have more of an ongoing, material presence in the places where we present them?

ROBBIE: I'd actually rather hear someone two generations down say, 'My grandfather told me how influential and positive the work of the Black Panther party was in Los Angeles'. What matters is that people in those places continue to tell the stories.

## Possibilities for social change

MARIE: For me, theatre can be both a way to conduct research and a way to communicate that research to a larger public, certainly beyond the limits of the scholarly journal article [see also Cahill and Torre, Chapter 23 in this volume]. In putting the process and results of research on stage, there's the possibility of participation that you don't necessarily get otherwise. Another thing we experienced with *Primary Sources* was how the media, most likely without intent, contributed to the participatory elements of the series, particularly in Boston. For example, *TURF* was on the front page of *The Boston Globe*, it was the subject of several National Public Radio [NPR] reports, and we were later able to produce an NPR version that was broadcast nationally with live call-ins in certain markets. All that effectively extended the dialogue. A newspaper article said, 'One thing is certain, people are talking'. With a participatory project, that's just what you want. And it's not just the participation of the people who made it.

ROBBIE: It's the participation of people we don't know; the performance provides an example, and it reflects more widely.

MARIE: One thing we're increasingly expected to do in academia is measure the impact of what we do [see also Cameron, Chapter 24 in this volume].

ROBBIE: I would rather think of a different kind of work in the academy, other than what we narrowly call academic. I found a surprising connection to other kinds of work when we were doing research for *Sugar* in Ohio.[2] As a result of being in story circles, people with diabetes promised to take better care of themselves. We now imagine that in the future we could measure if participants were influenced to be more rigorous about their diabetes management by exchanging stories and dialogue in PT. I am interested in how the commitment to speak stories may be measurable as a healing possibility.

MARIE: We certainly know that for some people with whom we worked, especially the actor-collaborators, participation in *Primary Sources* had long-term effects on their thoughts and actions.

ROBBIE: That's the most important thing. Remember Deborah in Mississippi? In the story circles, she was most resistant to the idea that there's anything more to know or change about race in this country. Then, during the time we were working together, she was in a store and noticed that a black man was being treated differently, and badly. She ended up telling that story in the performance.

MARIE: One of the things I remember is Tom from South Boston, and how the experience of being part of *TURF* and expressing group prejudice on stage was a release for him. It enabled him to come out as a gay man in his neighbourhood after living closeted there for years.

ROBBIE: When actor-collaborators spend time with us, calling up visions from community memories, they are able to generate connections with the wider community about charged subject matters that have been nervously silenced. When audiences see theatre and suspend disbelief for a concentrated period, they are often able to perceive more in 'real' situations.

MARIE: What about fear?

ROBBIE: Sometimes fear is fear! But in acting work, fear is often resistance, and resistance is information, so fear is … actor material. We talk about it, we tell stories about it, we name it; we sense it in one another. So the possibility of change is transmitted to audiences through the examples of listening and talking and confronting resistance within the actor-collaborators that they then make visible on stage. That is the hard work.

MARIE: And it has had its effect on people who are interested in speaking across differences within a community. Like at the South Boston performances of *TURF* – an African-American woman from Dorchester said during the talkback that she chose to see the piece there instead of in her neighbourhood because she wanted to take advantage of what she considered a rare opportunity to come to South Boston, to see a performance about race relations there, and then, potentially face-to-face, hear what a white person who had been involved in or had supported anti-bussing activities might have to say. And it seemed to be liberating for everyone.

## One more dialogue

ROBBIE: The visions of the work are short- and long-term. We have witnessed responses to the work that are immediately poetic, refreshingly emotional and reflect possibilities for social change.

MARIE: In the course of doing this work, were there some particular surprises that moved it forward?

ROBBIE: The one I think of is Kent telling that story in Jackson, Mississippi, about the 'blue-gum niggers',[3] how that became one of the jokes, one of the serious and releasing jokes of Mississippi Freedom, how absolutely excruciating it was for him to tell that story, and how receiving of it all the cast members were, particularly the African-Americans.

MARIE: Yeah, it gives me goose bumps …

ROBBIE: And who could have made that up? I mean, nobody could have made up the response to it. The story itself was hard enough …

MARIE: A sigh of relief went up from the audience.

ROBBIE: I think they got it: that transformation could happen to this mature man who had been carrying this racist story around with him all these years, and that the African-American members of the cast had sympathy for him.

MARIE: The self-proclaimed 'Mississippi redneck'.

ROBBIE: Yeah, who was telling the charged and previously unspeakable story of how he was taught to be racist.

## Conclusion

Public officials and others often state, 'We don't know how to talk about race'. Yet theatre based in dialogue allows ranges of feelings and thoughts to breathe among different people. We have found that at best talking is healing and at least it can shift views beautifully. Many have asked us if we consider our work political theatre. Of course it is, since we certainly have points of view about the subject matter. Also, we trust that interactions fostered through the work, involving people from diverse backgrounds in particular places, provide examples for useful discourse around difficult subjects. We like using theatre, while cultivating its aesthetics, to share research and to extend awareness of history and charged social dynamics to a large and targeted public. We have also witnessed its efficacy in continuing difficult conversations and in drawing individuals toward personal change. We continue to be impressed by the effect PT has on interviewees, actor-collaborators and audiences. The relief many have expressed to us is gratifying, whether they wanted their stories used or not, and measurable only in the warmth of the exchange. We remain interested in shaping and framing atmospheres for sharing stories and dialogues and find a PAR approach helpful. When collaborators and audiences can come together and be struck differently by the same charged material, our work is successful. A final stunning example is an on-stage reminiscence by Ona, an actor-collaborator playing herself in *Mississippi Freedom*: 'I travel a lot, and whenever I fly back into Mississippi, I look down and see the trees, the red clay hills, the church steeples, and I think "Oh, how beautiful, how beautiful"... and then I remember the horrors.'

Almost always greeted with silence, then thunderous applause, Ona's statement encapsulates the poetry and the promise that PT offers to those interested in taking up its challenges.

## Notes

1 *Mississippi Freedom* was performed in six locations in Mississippi and in modified form in Houston, TX, and New York, NY. We presented *TURF: A Conversational Concert in Black and White* in the Boston neighbourhoods of the South End, Charlestown, Dorchester and South Boston; a radio version was aired on dozens of National Public Radio stations across the US with live call-ins in several markets. *The Other Weapon* was performed in the South Central, Inglewood and Westwood sections of Los Angeles as well as in Hollywood.

2 Our most recent collaboration is *Sugar*, a work-in-progress dealing with complex issues of life and death surrounding the prevalence of diabetes within communities of colour. We worked on the piece in collaboration with local residents during a residency on the Ohio State University campus and in the surrounding Columbus, OH, area in January 2006.

3 As an improvisation, Kent told the story of how he and his friends as teenagers would cruise through town looking for dark-skinned black youth, whom they called 'blue-gummed niggers', to beat up. Kent said light-skinned blacks, or 'red-gummed niggers,' were less frequently their targets.

# 18 Photovoice

## Insights into marginalisation through a 'community lens' in Saskatchewan, Canada

*Brigette Krieg and Lana Roberts*

*Plate 18.1* 'The voice I found slung 'round my shoulders': participant-researcher with camera, July 2006

**Box 18.1 Telling our stories**

Through our own photography, we are given the opportunity to tell our stories in pictures and narratives. When a camera is placed in the hands of a marginalised people, whether it be by unemployment, poverty, homelessness, living with illness or facing any type of hardship, the opportunity is there to show others what life is really like in our own surroundings. Active participation also provides a venue where people can share their own ideas, views and experiences.

Source: Lana Roberts, Photographer/Researcher

## Introduction

Marginalisation refers to 'the context in which those who routinely experience inequality, injustice, and exploitation live their lives' (Brown and Strega 2005: 6). Further, marginalisation is indicative of experiences of injustice, inequality and exploitation, as well as lack of access to resources or power to create necessary change (Brown and Strega 2005). Examination of the issues requires a research process and methods that remain faithful to anti-oppressive practice and emphasise community involvement and ownership; where the end result of the process is not simply knowledge generation but a community response to a community issue (Potts and Brown 2005).

Photovoice is a Participatory Action Research (PAR) method that enables local people to identify and assess the strengths and concerns of their community (Wang 1999; Wang and Burris 1997). It has emerged as a potential tool for advancing knowledge around marginalisation. In this chapter, we reflect on our experiences of an ongoing Photovoice project in Prince Albert, Saskatchewan, Canada to argue that Photovoice is an appropriate tool to effectively and respectfully access the viewpoints of marginalised populations about their lived experiences of oppression and resistance. We present our perspectives through a dialogue between the researcher and a Photovoice participant-photographer using the interplay of the main body of the text (the researcher) and the text boxes (the photographer).

## Prince Albert Photovoice

The Prince Albert Photovoice process was precipitated by an event that occurred on 10 June 2005. Approximately 400 community members in Prince Albert, Saskatchewan came together to march and protest in response to Amnesty International's call for action against the marginalisation of Indigenous women throughout the world. The residents marched throughout the city and stopped at various locations to listen to the stories of family members either looking for missing women, or paying respects to those who had fallen victim to societal injustice. At almost every location were pre-posted photographs of the women which made their family members' stories of marginalisation much more powerful.

Amnesty International's (2004) paper *Stolen Sisters: A Human Rights Response to Discrimination and Violence against Indigenous Women in Canada* calls for community-led responses to address the 'social and economic marginalisation of Indigenous women' [that has consequently forced a] 'disproportionate number of Indigenous women into dangerous situations that include extreme poverty, homelessness, and prostitution' (2004: 2). In response to this call and for my doctoral dissertation, I initiated the Prince Albert Photovoice project. Its purpose was to share viewpoints of Indigenous women who experience discrimination and violence in their daily lives, through the use of visual images. Using Photovoice as a vehicle for change also enabled me to re-examine of the role of the researcher or 'expert' in community issues. The Photovoice participants – a group of multiply marginalised Indigenous women – made meaning of marginalisation and explored how that meaning informed their lives and communities.

## Photovoice: key elements

Photovoice is a grass roots community assessment tool that enables local people to identify, represent and enhance their community using photography as the medium for communication (Strack *et al.* 2004). Using photography as the catalyst for both individual and community change, Photovoice allows participants to document their own worlds, discuss issues with policy-makers and become active agents in social action (Wang and Burris 1997). Photovoice has been used to examine social and health issues of marginalised women, youth and homeless populations (McIntyre 2003; Strack *et al.* 2004; Wang 1999; Wang and Burris 1997).

### *Photovoice process*

Essential to the success of Photovoice are both the identification of the target audience and the selection of the participant base. As the outcome of Photovoice is to create an educational tool informed by local knowledge to influence public policy, identification of the target audience is key to the process. In the case of the Prince Albert Photovoice project, I (Brigette) was responsible for contacting potential partner organisations and making appointments to explain the process.

The immediate target audience for the project consisted of a small group of key decision makers who acted as an advisory committee to the Photovoice group and came to the table with the social and political desire to support the proposed changes identified by them (see also Wang 2003). The broader target audience included members of city hall, cultural centres, the local arts centre, social and health services' personnel, local policing authorities and all others who expressed interest in the project and its outcomes. Both the immediate and broader target audience members were invited to attend a community presentation and then request times and dates to have the photographers share their experiences with each organisation.

Wang (1999) recommends recruitment of seven to ten Photovoice participants as an ideal number for in-depth discussions. As this Photovoice project examined the defining characteristics and effects of multiple marginalisation of Indigenous women, participants were urban Aboriginal women in the city of Prince Albert, Saskatchewan who had experienced marginalisation, either directly or indirectly. The potential participants were invited to come and learn about the Photovoice process through word of mouth and public flyers. Neither I nor the community leaders recruited participants directly so as to avoid any immediate power imbalances. Each participant joined with their own reasons for becoming part of the project, as is evidenced in Lana's reflections in Box 18.2.

The prospective participants attended an initial meeting to discuss the Photovoice process, the role each participant would play in the project, the underlying issues around the use of the cameras, power and ethics, potential risks, and

**Box 18.2 Bringing out the strengths in our community**

First, I feel it is a privilege to have a part in this venture. Second, as an Indigenous woman I feel there are various issues in my community that need to be addressed. I would agree that a Photovoice project is an ideal technique. Some of the issues that I feel are prevalent within the community are: homelessness (for a high number of Indigenous women); poverty; unemployment (or lack of a proper income to live a decent or comfortable life); illness; and lack of education and decent wages. Of course, not to focus on that which is negative, this community also has a depth of strength and that is where a Photovoice project is beneficial in helping to bring out the strengths within the community. At the political level, I would like to see some of these issues being addressed, so that something positive would be done. Photovoice can get us there.

Source: Lana Roberts

how these could be minimised (see also Adams and Moore, Chapter 6 in this volume). The three central goals of the photographer's role were identified: their status as experts, their potential to educate and inform on important community issues and their ability through Photovoice to influence public policy (Wang 1999). Only when photographers were fully informed of the process, risks and benefits were they given the opportunity to decide about the degree of their participation and asked to read and sign consent forms (see also Manzo and Brightbill, Chapter 5 in this volume).

When the group of photographers solidified, I posed an initial theme of marginalisation for taking pictures, however through discussions the group identified further themes on key issues, how these issues were personally defined for each photographer and how they might be visually represented. Some of these themes included poverty, housing, transportation, worker–client relationships and racism. Discussions involved both strengths and weaknesses, brainstorming on possible solutions and the role photographers might play in igniting change at a local level (Wang and Burris 1997). The photographers were also given a journal to write down their ideas around the central theme of marginalisation, possible photographs to accompany them and instructions on the appropriate camera settings to obtain specific photographic outcomes.

The cameras were distributed and direction was given on how to use them. The use of a Holga medium format camera had been recommended by several Photovoice facilitators for its ability to layer pictures, distort images through special lenses and distort colour with colour filters. The Holga camera was also relatively inexpensive and replaceable if lost or stolen. It allowed the photographers to take creative pictures (see Plates 18.2–18.5, Box 18.3) that were perhaps more symbolic rather than literal representations of the project theme (Wang *et al.* 2004). I gave a date for film return to maintain motivation and provided support throughout the process (Wang 2003).

*Plate 18.2* (Credit: Lana Roberts 2006, with filter)

*Plate 18.3* (Credit: Lana Roberts 2006, layered photos)

When the photographers felt they had exhausted their photographic possibilities, the group met to discuss their photographs. Participants put a lot of time and energy into creating photographs that accurately represented the project's key issues and were involved in the entire three-phase process that provided the foundation for analysis (see Wang 2003).

The first phase – *selecting photographs* – was necessary to open the doors for discussion. First, photographers selected five to seven of their favourite photographs. Then each photographer told the story in the picture to other photographers and me, including why she chose the picture and any stories that may have been shared by the person being photographed (Lykes 2001a).

Wang (1999) suggests that the photographers frame stories around their photos using an approach called SHOwED, which allows the group to critically analyse the content of their photos through a series of questions: 'What do we See here? What is really Happening here? How does this relate to Our lives? Why does this situation, concern or strength Exist? What can we Do about it?' (Wang 1999: 188).[1]

Wang's (1999) approach to interpreting and analysing the content of the photographs has not always been successful. In her work with experiences of Irish women, McIntyre (2003) found the SHOwED approach limited the personal interpretation of the individual photographers. Instead, McIntyre (2003: 53) suggests that the photographers rely on instinct when choosing their photos and interpret and analyse photos with the following questions: 'What did the photographs mean to you? What is the relationship between the content of the photographs and how you perceived the community? How do you see the photographs as reflecting issues that are salient to you in this community?' Although

*Plate 18.4* (Credit: Lana Roberts 2006, layered with filter)

*Plate 18.5* (Credit: Lana Roberts 2006, with distortion lens)

the questions differ from Wang's (1999), the importance of dialogue is still emphasised to help people gain a clearer sense of the stories they want to accompany their photos (McIntyre 2003).

The second phase – *contextualising* – occurs through group discussion and sets the stage for voicing individual and group experience (Wang and Burris 1997). In the second round of analysis in our project, a small group chose two to four photographs and organised them into topic groups. Together the women clustered their ideas; identified similarities across photos and constructed a holistic analysis of the clusters of photographs through open dialogue (Lykes 2001a).

**Box 18.3 Seeing what life is really like for us**

The pictures taken can portray and reveal the reality of lived experiences faced daily by people in communities like mine. This form and process of documentation can help bring attention to various conditions such as homelessness, poverty, unsuitable housing and other multiple social factors. People on the outside can view how social issues affect the people who are working the cameras. This helps to reveal how the world of the marginalised is oftentimes much more different than those who are looking in imagine it to be. People who do not have a clue about some of the social conditions faced daily by people on the margins can see what life is really like for us. The people at the grassroots level are given a chance to prove and express themselves rather than have someone else speak for them.

Source: Lana Roberts

The process of attaching multiple meanings to a single image, lead to the next phase of the dialectic process – *codifying*. During this stage, three types of dimension emerged: issues, themes or theories (Wang and Burris 1997). From these, photographers identified achievable target actions including a format through which to share their images and stories with non-participants. The format can take the form of a slide show presentation, and/or an art exhibition at the local gallery, but key to the success of the approach is that it should be shared with an audience of policy-makers, journalists, researchers and community members (Strack *et al.* 2004; Wang 1999). In the case of our project, the final presentation will be shared initially with family and friends at a small dinner, as an art opening at the local art gallery and then at an international conference. As well, the final project will be available on the Internet and registered with the national and international Photovoice database.

Like all good PAR, it is essential that participants 'own' the project and control its outcomes. In essence, this Photovoice process enabled participants who had dedicated their time to exploring and understanding the issues and effects of marginalisation to meet with a group of key community decision makers, and through their photos and presentations, to challenge the existing stereotypes that these individuals and the organisations they represent help to perpetuate. In so doing, Photovoice gave this marginalised population a voice in political arenas that had the potential to catalyse political and societal change by circulating new and positive images of them (McIntyre 2003; Wang 1999).

## Learning along the way: reflecting on the Photovoice process

Photovoice in Canada is in its infancy with only a small handful of high quality examples to draw upon. That said, within this context, the immediacy and power of the visual images have created a platform for dialogue and mutual learning on pertinent issues (Box 18.4).

### *Strengths*

Perhaps one of the most essential benefits of the Photovoice process in our project was its ability to stimulate discussion and motivation for social action in our community. Photovoice proved to be an effective means of shifting the local power balance, encouraging ordinary people to become advocates in their own reality.

In other cases, Photovoice has also motivated groups to action. In a project on women in poverty, Women in Fair Income, Scott (2005) noted a 'domino effect' in which her Photovoice project galvanised a whole range of other initiatives. In this case, initial response to the photographs acted as impetus for similar projects conducted by the Prairie Women's Health Centre of Excellence (PWHCE). In response, PWHCE went on to conduct similar projects in Alberta and Saskatchewan. The first in Winnipeg, entitled Poverty: Our Voices, Our Views and another in Saskatoon entitled, Looking out/Looking in: Women, Poverty, and Public

**Box 18.4 Encouragement to act**

This particular type of research is proactive. Photovoice brings the grass-roots community together, and helps to form a collaborative effort between various agencies, which helps make positive changes where needed. Personally, I see this as a stepping stone for many of us to work towards betterment in our community and bring out into the open many of the hidden talents, strengths and knowledge of the Indigenous women. Especially for those who do not realise what strengths they have, until someone has placed that grain of encouragement and hope into believing that they, as marginalised individuals, groups or communities, have a voice that should be heard. As the cameras are placed into participants' hands, I for one believe that many individuals will once again begin to see how they can be involved in working towards change. Through this, we are given hope and empowerment to do something about our own lives.

Source: Lana Roberts

Policy depicted women's experiences of living in poverty, the barriers and the supports found in each community, as well as critical examination of the policies and programmes available (Willson 2006).

Importantly, Photovoice helped to rescale the problems faced by individual communities by offering a standpoint from which to compare and contrast experiences of poverty from three separate yet interconnected viewpoints. It provided a means for marginalised communities to connect and learn from each other, then to scale up their responses and present a broader front to key decision makers. In each project the primary presenters were the photographers who shared their photos during community displays and led discussions about the issues related to poverty. The strength of the individual presenters and the group as a whole was very apparent to the audience members, demonstrating the potential to empower participants through enhanced self-esteem and increased political efficacy (see also Wang 1999; and Pratt *et al.*, Chapter 12 in this volume).

Further, our project presented the community, quite literally, through a new lens; one that no longer pathologised them as deficient but as strong and possessing the capacity to take action. Photovoice helped in documenting the successes and failures of previous community activities and interventions. This allowed the community the opportunity to assess policy and programmes affecting them and provide supporting visual evidence (see also St. Martin and Hall-Arber, Chapter 7 in this volume; Wang 1999).

### *Challenges*

Like many aspects of PAR, establishing a Photovoice project requires a major time commitment from both the researcher and photographers (see also Hume-Cook *et al.*, Chapter 19 in this volume). They work together to create the end products that will eventually become a community education tool delivered by the photographers. The researcher must take responsibility for the challenging aspects of completing foundational legwork: getting ethics approval, establishing a target audience, organising participants, facilitating meetings and accommodating the needs of the participants. However, the participants must also invest a lot of time, which, depending on the type of participant, can be challenging.

Participants in the Prince Albert project are single mothers, some living in poverty, some without reliable transportation and many grappling with multiple demands on their time and resources. To accommodate these barriers, we identified a convenient night of the week and a central location, and treated each Photovoice session as a gathering, providing transportation, a shared meal and childcare to enable parents to participate.

The power of Photovoice lies in its ability to make direct connections between locally identified community concerns and municipal, provincial and national policies. Change is a relatively attainable goal at the local level but sharing knowledge and organising action provincially, nationally and internationally can be an expensive endeavour. The financial burdens of running a Photovoice project are possibly one of the biggest challenges. Cameras, film and developing alone require substantial financial resources and the cost of arranging travel and accommodation for photographers to share knowledge can be prohibitive.

This financial responsibility translates into issues of access to resources and power when deciding who will be part of the dissemination of findings. The Prince Albert Photovoice project had initially stagnated because I was covering all costs associated with running the project. However, it has recently regained momentum through a community based research grant from the Indigenous People's Health Research Centre.

Finally, because Photovoice is a relatively new research method, documentation of the process is limited and requires new users to progress by trial and error. There are many other challenges (Box 18.5).

What we have learnt is that discussions must include attention to the potential personal effects of participating in the Photovoice process. Critical examination of marginalisation may cause negative emotional responses because of the

**Box 18.5 Challenges**

Some challenges I see are: convincing people to become involved, and gaining the trust of the community; bureaucratic opposition (certain people not agreeing or being enthused about such a project); and people such as transients or the homeless, not fully cooperating, or not remaining committed until the end of the project.

Source: Lana Roberts

subject matter or from not getting the desired response from the target audience. Training must therefore include examination of the possibility of unexpected outcomes (see also Adams and Moore, Chapter 6 in this volume; Wang 1999; Wang and Burris 1997).

## Conclusion

Using Photovoice to address the issues of marginalisation demonstrates how a PAR process can facilitate people to use their knowledge and abilities to effect necessary changes at a variety of scales. Photovoice emphasises the importance of community involvement and ownership, and the need for the end result of the process not simply to be knowledge for its own sake but action on community issues. However, Photovoice is in its infancy and to develop it as an anti-oppressive research method, information and experiences need to be documented and shared. Further, remaining faithful to the emancipatory tradition of this process means that future documentation should be inclusive of both researcher and participant perspectives.

## Notes

1 These stages are repeated for each roll of film that the photographers have to share with the group until a series of themes around the issues emerge.

# 19 Uniting people with place using participatory video in Aotearoa/New Zealand

## A *Ngāti Hauiti* journey

*Geoffrey Hume-Cook, Thomas Curtis, Kirsty Woods, Joyce Potaka, Adrian Tangaroa Wagner and Sara Kindon*

### Introduction

*Piki ake ki te Taumata o Mekura ko Ruahine*
*Titiro atu ki te maunga tapu Aorangi*
*Heke iho ki te awa e rere nei ko Rangitīkei*
*Ka pari a ki uta ki Pātea*
*Ka pari a ki uta ki Otoea*
*Ka pari a ki uta ki Otara*
*Ka tatū ki Te Hou Hou nei*
*Ko Ngāti Hauiti e tū ake nei*

*Ki ngā tini aituâ haere ki tua o te ārai*
*Moe mai i raro i te parirau o te Kai-hanga*
*Heoi anō!*

*E ngā mana, e ngā reo, e ngā waka*
*Ka nui te mihi, ka nui te aroha*
*Tēnā koutou, tēnā koutou, tēnā koutou katoa.*

Ascend the summit of Mekura on Ruahine
Gaze yonder to the sanctified mountain of Aorangi
Descend upon the currents of the Rangitīkei
To settle at Pātea, Otoea and Otara
Finally reaching Te Hou Hou
Tis Ngāti Hauiti standing before you

To the dearly departed, go beyond the veil of darkness
And sleep beneath the wing of our Creator
And moving on …

To the pillars of society, the voices of influence and to the nation
Greetings, salutations and affection to one and all.[1]

In this chapter, we discuss a collaborative research partnership, which has been using Participatory Video (PV) to explore relationships between place, identity and social cohesion in Aotearoa/New Zealand. We comprise an Australian independent audio-video producer based in Wellington (Geoff); members of *Ngāti Hauiti*, a Māori 'tribe' in the lower North Island (Thomas, Kirsty, Joyce and Adrian); and an English permanent resident geographer working at Victoria University of Wellington (Sara). Our collaborative research has been informed by Participatory Action Research (PAR), critical ethnographic media practices (Pink 2007), and Kaupapa Māori research, which argues for research by, with and for Māori (Smith 1999).

Introducing the background and common processes associated with PV from its late 1960s inception, to its present day use, we then outline how PV was used within our project. We note here that our orientation to PV was associated with community research and development rather than more activist interventions (Gregory 2005; cf Chatterton *et al.*, Chapter 25 in this volume). We point to the challenges we encountered and reflect on some of the outcomes we achieved. Finally, we offer some parting thoughts about lessons learned along our journey, and offer them here in the spirit of *manākitanga* (blessings) and *aroha* (compassion).

## History and evolution of participatory video

> Participatory video is a special kind of storytelling that ideally involves the community in telling a story, listening to the story, interpreting the story through its own lens and being empowered to retell and change it to create a community – a political reality –- that matches one's desired condition
>
> Bery 2003: 102

The first documented participatory use of video occurred on Fogo Island in Newfoundland, Canada in the late 1960s and lasted for eighteen months (Crocker 2003). Widely referred to as the Fogo Experiment, the pioneering interactive use of video as a tool for social change was revolutionary in its desire not just to document social problems, but also to use the audio-visual process to raise awareness and share information within and beyond the communities involved.

By putting community members behind as well as in front of the camera and by facilitating a process of community feedback on the films produced, research participants became 'meaning makers' who explored and worked to change their own 'realities' through the production and analysis of video products (Crocker 2003). The process enabled participants' technical empowerment (Guidi 2003) by increasing their knowledge and confidence using video and editing equipment. It also enabled their personal and community empowerment through seeing their lives on screen and discussing possible actions for change (Crocker 2003; Stuart and Bery 1996).

Over the last twenty years, PV has become more widely integrated into development practice (Lunch and Lunch 2006). It is used to support community education, cultural identity and preservation, and to inform political participation and social change (Braden 1998; Mayer 2000; Van Vlaenderen 1999). It is an effective tool for Participatory Research in community development (Braden and Mayo 1999; Frost and Jones 1998). At the heart of PV is the commitment that participants create videos according to their own priorities; they control how they are represented and they decide what the video will be used for. This differs from most examples of documentary film where outsider video producers tend to retain control.[2]

## Process

By definition, Participatory Video is not beholden to commercial television or film imperatives. Rather, the primary outcome desired is 'that of the interaction of individuals and their own personal growth that comes about during the process of production' (Guidi 2003: 263). In many cases, the sense of community and cooperation involved in making a film is paramount (Crocker 2003). In others, the ability to influence decisions affecting their lives is a key motivation. A typical process might involve the stages outlined in Box 19.1.

With these stages in mind, it is important to note that PV, like other participatory communication methods, can and does take many forms because it has to be adapted to suit the always politicised context and timeframe within which it is being used (Servaes 1996). Stages can last between a few days to several years depending on the purpose of the initiative, the degree of social change desired, the level of funding and the nature of the relationship between outside researchers and/or video producers and community members.

### Box 19.1 Typical stages in a Participatory Video process

1. Contact and establishment of relationship between outsiders (academic and video producers) and community members.
2. Negotiation of purpose, process and outcomes (this may include a Memorandum of Understanding covering copyright/ownership of media produced).
3. Identification of, and negotiation with, participants to be trained.
4. Negotiation of access to equipment.
5. Training and familiarity with equipment, concepts and techniques.
6. Work taken to wider community and discussed.
7. Possible change or action as a result of discussion.
8. Management of products (dissemination, storage, access).
9. Motivation to create more projects of benefit to the community.
10. Liaison with outsiders about future and ongoing involvement.

Source: Adapted from Guidi 2003 and authors' own experiences

## Participatory Video in Te Whakaohotanga O Ngāti Hauiti

In 1997 Sara asked *Te Rūnanga o Ngāti Hauiti* (the Tribal Council of the Hauiti People) to consider becoming involved in her university-funded project using PAR and video. The project's academic goal was to explore relationships between place, identity and social cohesion within remote rural communities in Aotearoa/New Zealand.[3] The Rūnanga considered the request and delegated a subsidiary entity within the *iwi* (tribe) to manage the process. As Thomas, Chairperson of the Potaka Whānau Trust, recollects:

> My introduction to the research project was at our *whānau* (family) land trust meeting. I gave the proposal a once over before our meeting and thought to myself that we would need to ensure that *Ngāti Hauiti* interests were protected. After the meeting we set aside some time to talk with Sara and Geoff. It wasn't long before I could see this would be a very valuable project for *Ngāti Hauiti*.

A Working Party from the Trust (Thomas, Kirsty and two other members of *Ngāti Hauiti*) was established to oversee the project. Geoff and Sara met regularly with them during 1998 to establish the project's principles, aims and design. The ownership of any results (transcripts, videos and publications produced) was negotiated (see also Manzo and Brightbill, Chapter 5 in this volume). The Working Party named the project: *Te Whakaohotanga o Ngāti Hauiti* (The Awakening of *Ngāti Hauiti*). Thomas notes:

> We also used a saying *Mahi Nga Tahi* (working together as one) to inform our process. We documented all our decisions in a Memorandum of Understanding (MoU) in January 1999, before the start of community-based research with other members of the *iwi*.[4]

Concurrent with the development of the research focus and the MoU, Geoff and Sara carried out video and research training with a group of between seven and fifteen people. Seven formed a core Community Video Research Team (CVRT). Geoff recounts:

> They participated in fifteen sessions (about sixty hours) of video production training, which covered audio-visual language and concepts, camera skills, lighting, sound, direction and editing. All sessions involved hands-on group work. We engaged with the medium from mainstream and critical perspectives (i.e., not just 'this is how it's done', but considering 'the reasons why it's done a certain way, the effects this has, and how it can be done in other ways, with different effects'). Every session of training was taped for people missing individual sessions and for archival purposes.

The CVRT also had three sessions (about nine hours) of training with Sara, which used participatory exercises and role plays to build skills in community research, ethics and interviewing (see Plate 19.1). As she notes:

> In the participatory exercises, CVRT members shared information about *iwi* members' relationships with each other and the *rohe* (tribal domain). They also developed skills they could use later in their own tribal research.

Given the project's ostensive focus on relationships between place, identity and social cohesion, the CVRT brainstormed topics they wanted to pursue, and that would produce interesting information for others in the *iwi*, as they refined their skills.[5] From the list of topics generated, they decided to make their first training video on *wāhi tapu* (sacred places) in the *rohe*, which they set to a soundtrack of one of the *iwi waiata* (songs/chants).

After the *waiata* video was completed, the *Rūnanga* requested that the CVRT carry out video interviews with *iwi* members living in Utiku township to find out more about their relationship to it, the *rohe* more broadly, and their participation or otherwise in *iwi* affairs. Joyce and another CVRT member, Raihania Potaka, organised and carried these out with support from Geoff and Sara (see Plate 19.2).

Just after the video interviews was a major *iwi* event within the *rohe*: the inaugural five-day *Awa Hikoi* (river journey), along the most significant *awa* (river) of the *iwi* – the Rangitīkei. It involved twenty-five *iwi* members from around the country in activities both on and off the *awa*, including the study of *whakapapa* (genealogy) and *wāhi tapu*. Kirsty comments:

*Plate 19.1* Brainstorming possible whānau members to interview. Left to right: Joyce Potaka; Reihania Potaka (standing); Harry Lomax; Sara Kindon; and Arnold Potaka (Credit: Geoffrey Hume-Cook 1999)

*Plate 19.2* Video interview in Utiku Township. Left to right: Sara Kindon (interviewer); Cookie Bennett (Utiku township resident and *iwi* member); and Joyce Potaka (videographer). (Credit: Geoffrey Hume-Cook 1999)

> The river used to be the main 'highway' through our region. The aim of the journey was to encourage *whānau* (family) members to visit important places along the way, share stories about our history and get to know each other.

While not part of the original research activities negotiated with the Working Party and CVRT, the *Rūnanga* authorised documentation of the event.

Thomas reflects:

> This was a memorable trip. Experiences we were able to capture included some very touching *kōrero* (talk/speech) from a revered *kuia* (female elder) who has now passed on.

Between 2001 and 2003, Thomas, Kirsty, Geoff and Sara conducted oral history interviews with *Awa Hikoi* participants to create an archive and gather additional audio for a collaboratively produced, feature-length documentary of the event. Since 2001, Adrian has become involved in collaborative writing, presentations, and the *Awa Hikoi* documentary.

## Challenges and outcomes

First, PV within the context of PAR requires considerable voluntary time commitment. This was difficult for the small numbers of people in the *rohe* who were already involved in other *iwi* activities or needed to take up seasonal work. It was impossible to maintain a consistent group of *iwi* researchers in the CVRT.

Second, as with other technologically 'complicated' adjuncts to PAR (i.e., GIS, see Elwood *et al.*, Chapter 20 in this volume), Geoff faced a number of delicate balancing acts: one was between facilitating CVRT familiarity with some complex audio-visual techniques and technology in order to effectively tell their stories, and 'delivering' tangible outcomes for the *iwi* and the funding institution within a timeline.

Another was dealing with television's ubiquity: all creative work involves choices, but choices about re-presenting life-worlds using a medium that is taken for granted asks participants to reassess assumptions about their world. This involved the raising of media awareness (for example, demystifying 'realist' cinematic conventions) whilst simultaneously encouraging the sense of possibility the medium allows. Also, given the CVRT members' exposure to mainstream television, Geoff needed to ensure that they understood the 'production values' context – high-budget broadcast versus low-budget research/PV– of the overall project. This was important so that the difference between a (possibly hoped for) 'mainstream' outcome and their PV one would not be judged harshly by comparison.

Third, it was not until actually faced with 'editorial' decisions in the first PV product (the *waiata* video) that cultural understandings and ways of seeing, came to the fore. A shot taken by one of the CVRT members was a tilt-up from the base of a carving. When the rough cut of the *waiata* video was taken to the *kaumātua* (male and female elders) on the *Rūnanga* for feedback, they pointed out that this directionality alone was an inappropriate act: in effect, disrespecting an ancestor in the carving. Fortunately, reversing the shot in post-production was a simple task and the final edit respected all *Ngāti Hauiti tikanga* (protocols) and *kawa* (conventions).

Fourth, there is increasing concern (Kesby 2007a) about the ability of PAR to have an effect beyond the project context. PV would appear to offer potential here, yet in our experience, the video products' ability to travel into different spaces and reach different audiences should not be assumed (see also Cahill and Torre, Chapter 23 in this volume). For example, the CVRT's *waiata* video was shown, in conjunction with a short video of the training process edited by Geoff, to two mostly academic audiences with very different responses (Box 19.2).

Alongside these challenges, we have achieved outcomes of which we are proud. We have an archive of audio-visual materials about the relationships between place, identity and social cohesion within *Ngāti Hauiti* at the turn of the millennium. This serves as a valuable resource for future generations and for our ongoing collaboration. The *Awa Hikoi* documentary is almost complete and will support the desire of the *iwi* to re-unite people with the *rohe*. Thomas notes: 'as a result of participating in this project, we have been able to make more conscious choices about how we use video and what we film.'

At a more personal level, we have all benefited from our association with this project in terms of the technical skills learned, the personal relationships developed and the support that we have shared outside of the project context in other areas of our lives and work. Key CVRT member, Joyce, reflects below (Box 19.3).

### Box 19.2 Considering audiences

In one instance, we participated in a seminar with a group of university lecturers and students. One lecturer described the *waiata* video as 'just another family video' – suggesting it had no academic significance. He seemed to miss the point about the process that created the video and what it meant for those involved.

In another, we were challenged by a prominent *Māori* film maker who insinuated that his approach to film making was authentic, whereas what we were doing was not. Part of his criticism targeted Geoff's use (in the training video) of the story board as a 'colonising process'. The fact that he did not wait to receive a response from me as the *iwi* representative, and wasn't prepared to discuss an alternative view, only confirmed in my mind that he was challenged by what we were doing. After all – it is for *Ngāti Hauti* to decide what's appropriate for us, not him.

Source: Kirsty Woods March 2006

### Box 19.3 More choices

I was a mum at home with kids and a young baby and was a bit house-bound. I had a routine of *kōhanga* (*Māori* language preschool), dinner, watching TV. So to go and do a creative exercise was an opportunity [. . .] and it lead to a lot of interesting things for me. [. . .]

I didn't know a lot of my family, my own history or anything like that, as I was adopted as a child by *Pākehā* (European New Zealanders). So it gave me an opportunity – just boots and all, straight in [. . .] We learned right from the beginning: from story boarding to lighting, proper camera shots and technical knowledge.

I wouldn't have had another opportunity to learn those sorts of things. I found for myself, well six years down the track, that this lead to other things for me. It lead to me doing a diploma at *Te Wānanga o Raukawa* (Raukawa Tertiary Institution) . . . . That was because of my participation in this type of research. So I feel that this sort of research does have a lot to offer and that you take what you want from it, you take from it what you feel that you can use. [. . .] And that's where I found myself. Being able to pick and choose and I've now got more choices in my life than I had before.

Source: Joyce Potaka July 2005

**Box 19.4 Lessons learned working with Participatory Video**

1 It's entirely possible to do research like this, even if you're not an experienced video producer or academic researcher.
2 Quality, affordable technology is available so buy/hire low-end professional camera(s), but don't forget good peripherals (tripods, lights and audio).
3 Despite the ease of access to technology and the ubiquity of the medium, it's still vital to get *au fait* with conventional techniques. Time needs to be allocated for this with a compatible trainer. (For advice and how-to guidance in the absence of suitable trainers, see Lunch & Lunch 2006 or Gregory *et al.* 2005).
4 *Arohanui* (Take time to understand and value the people involved and their contributions.)
5 Reaching a point of acceptance within ourselves as research collaborators takes time, shared space and history.
6 Being open to non-academic ways of writing and presenting works well to integrate various research collaborators' and participants' perspectives, and to reach a wide range of audiences.
7 Choose audiences carefully – know what space (hostile, nurturing) you may be stepping into before sharing video products.

Source: Authors' own experiences

## Parting thoughts

We have learned many things throughout this process. In Box 19.4 we outline seven points that seem significant to us and might help others considering PV in other places.

Finally, while we have written this chapter together, we would like to close with some words from Thomas, which we feel speak to the heart of our relationship and experiences.

> The destination is the goal, but the journey is the gold. I liken it to walking the Great Wall of China [...] The destination is known, the method of transport is walking, but the journey is the treasure. In this project, the method of transport is Participatory Video and the *taonga* (treasures) are the experiences that we capture along the way.
>
> *Nāu te raurau,*
> *Nāku te raurau,*
> *Ka ora ai ngā tāngata!*
>
> With your contribution
> and my contribution,
> together we will get there!

## Notes

1 Official *korero* (speeches) begin with a *mihi* (a formal speaking structure used during welcoming). It positions the speaker in place, time and genealogy. This *Ngāti Hauiti mihi* provides the foundation for our chapter. For more information see www.ngatihauiti.iwi.nz.

2 With the exception of consciously participatory or indigenous ethnographic film (see Pink 2007).

3 A relatively small *iwi*, in the 2006 national census, 1038 people identified officially as *Ngāti* Hauiti.

4 At the end of the project, copies of all video footage, research reports and academic papers will be lodged with *Te Rūnanga o Ngāti Hauiti*.

5 We use the term 'ostensive focus' because, as an academically funded project, the 'research paradigm' applicable in social sciences anticipates 'an' outcome, one that is directly related to a research 'question'. As PAR, the collaborative defining of the research outcome did, fortunately, fall within the broad parameters defined in the initial project proposal.

# 20 Participatory GIS

## The Humboldt/West Humboldt Park Community GIS Project, Chicago, USA[1]

*Sarah Elwood, Ruben Feliciano, Kathleen Gems, Nandhini Gulasingam, William Howard, Reid Mackin, Eliud Medina, Niuris Ramos and Sobeida Sierra*

### Introduction

Participatory GIS (PGIS) is an intersection of academic and activist practice that emerged in response to criticisms of GIS. Geographic Information Systems (GIS) are software that enable mapping and data analysis and are used in a wide range of circumstances, including urban planning, natural resource management, emergency response and community activism. Critics have been concerned that the high financial, time and training requirements of GIS can exclude grass roots groups, and that GIS may not be able to represent diverse forms of local knowledge (Elwood and Leitner 1998; Sheppard 1995). To try to address these concerns, PGIS initiatives typically work to adapt GIS software to better incorporate local expertise and knowledge, or use GIS within a Participatory Action Research (PAR) framework (Craig et *al.* 2002; Sieber 2004).

As a kind of PAR, PGIS has unique challenges and opportunities. Creating an inclusive research process can be difficult when a core component of the methodology is a resource-intensive computer technology that requires specialised skills (see also Hume-Cook *et al.*, Chapter 19 in this volume). As well, GIS has difficulty incorporating non-cartographic information about space and place. But for the 'action' part of PAR, GIS is a uniquely powerful and flexible resource (see also Chatterton *et al.*, Chapter 25 in this volume). It can be used to create diverse visual representations of place, community, needs and assets. These visualisations are especially influential with policy-makers, funding agencies, residents and other organisations.

A great deal has been written about these and other advantages and disadvantages of PGIS, but most of the existing accounts of the impacts have been authored by academic partners. In this chapter we provide a collectively authored discussion of our use of GIS as a negotiating tool, and lessons we have learned about sustaining university-community PGIS projects. Our author group includes a university-based researcher (Sarah), the executive directors of the Near Northwest Neighbourhood

Network (NNNN) and the West Humboldt Park Family and Community Development Council (the 'Development Council') (Bill and Eliud), a university-based research assistant (Nandhini), and past and present staff members of NNNN and the Development Council (Kate, Reid, Niuris, Lily and Ruben).

In 2003, NNNN, the Development Council and Sarah initiated the Humboldt Park/West Humboldt Park Community GIS Project (HPCGIS), a long-term research, education and community capacity building project in Chicago, USA. The HPCGIS project is a kind of PGIS project intended to develop strategies for sustaining GIS capacity in community organisations, collaborative learning opportunities between university students and community organisations, and to enable a better understanding of how GIS use affects community-based organisations' independence and power in neighbourhood redevelopment. We hope to create lasting resources that the NNNN, the Development Council and other organisations in the Chicago's Humboldt Park area can use in community-controlled redevelopment.[2] NNNN and the Development Council are non-profit community organisations whose activities range from affordable housing and economic development, to safety and emergency planning with residents, public school reform, and youth and family programmes. Local residents' concerns include affordable housing, public school problems, and higher levels of crime, poverty and unemployment than other parts of the city. We also have tremendous assets, including residents who are highly knowledgeable about community needs

*Plate 20.1* HPCGIS 2005 Summer Workshop, using GPS for field data collection (Credit: Sarah Elwood 2005)

and conditions, a strong coalition of active non-profit organisations (the Humboldt Park Empowerment Partnership or 'HPEP'), and experienced activists and community development professionals.

The HPCGIS project involves several types of activities. With the support of grant funding, we established GIS 'mini-labs' at the NNNN and Development Council offices, and created a spatial data library that the NNNN and the Development Council use for their mapping activities. The library includes data from government sources and locally-collected information about community services, land use or property conditions. University GIS students work with NNNN and Development Council staff members on new data acquisition or mapping efforts. At weekly work sessions held at their offices, a research assistant (the community coordinator) helps with software or data problems, advises on new mapping projects, and helps new participants from the NNNN, the Development Council or other local community organisations learn GIS skills. Twice a year, we hold multi-day workshops for all participants to work on small GIS projects useful to both groups (Plate 20.1). Finally, participating staff members from the NNNN and the Development Council meet with Sarah, the university researcher, for interviews about the use and impacts of data and maps produced.

## GIS as negotiating technology

GIS is most influential for our planning, redevelopment and advocacy work when we use it as a negotiating tool to complement our existing knowledge and activities. GIS helps us create powerful visual messages about the community, its needs and our priorities. GIS is a tool to complement our mission and our goals, conveying information to residents, funding agencies and our elected officials. This technical capacity can have a significant impact on our efforts to advocate for the community. But the power of GIS depends on more than technology or technical expertise. The maps we make draw on deep knowledge and close observation of local conditions, and this knowledge strengthens our negotiations with government officials, potential funding agencies, residents and other community organisations. In community planning, if we make a map, it is important to get out and visit the site to get a three dimensional sense of it, before using the computer to add elements like air photos, parcels, or property identification numbers. Computer truth is not always 'truth', so it is critical to complement the GIS on the ground.

Our GIS activities are most influential when we gather our own information and use it together with local government data. For instance, the Development Council staff created the map in Figure 20.1 for work with several partner organisations, to try to identify a site for a community centre with after-school programmes. We began by making a map of existing community centres in West Humboldt Park. At first there appeared to be several community centres near our target area, but later we added our own information about programmes offered and age groups served. We found food pantries, shelters, and locations that only help seniors, plus religious centres, but nothing for youth. This further information strengthened our proposal, showing that we needed a community centre serving teens and young

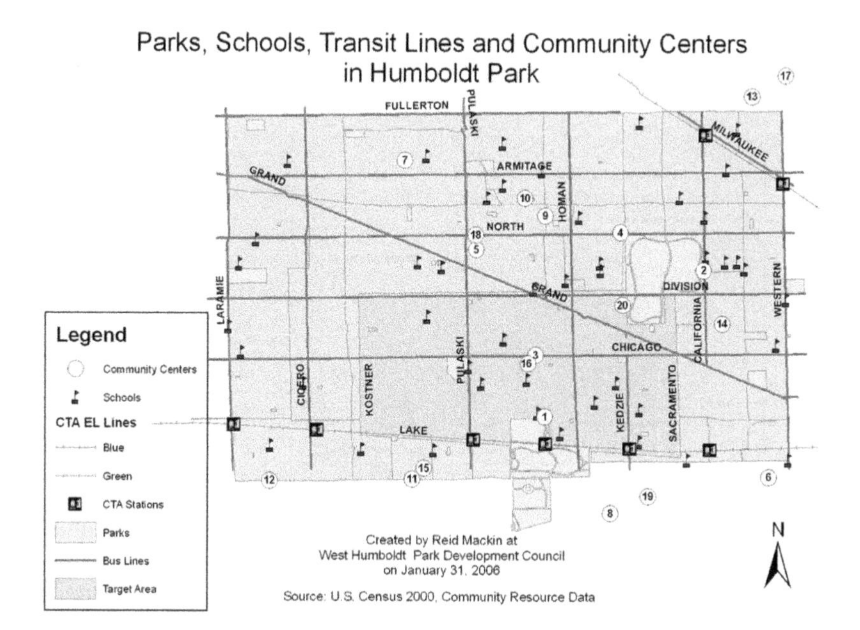


*Figure 20.1* Parks, schools, transit lines and community centres in Humboldt Park. Created by Reid Mackin at West Humboldt Park Development Council, January 2006 (Source: US Census 2000, Community Resource Data)

adults. Being able to illustrate this with our own maps helped us be more influential in negotiating with local government and other organisations, because we could show specifically what was there and whose needs were not being met (see also St. Martin and Hall-Arber, Chapter 7; Gavin *et al.*, Chapter 8 in this volume).

We use our GIS-based maps in flexible ways to communicate with City officials, funders, residents or other community organisations. A map needs to tell a story. If we can tell several stories with the same map, we can negotiate more widely for resources, with more potential allies. Figure 20.2, a map by the NNNN, has been used this way. It shows past and present affordable housing and economic development projects, and significant cultural sites and community resources. We can use it to show what the NNNN has accomplished; when we take potential funding agencies on tours of the community, we hand out this map. It also tells us how our own work is balanced across the redevelopment area, as we plan our own activities. The map can also tell a story about community needs, illustrating to local government that we still need more affordable housing. Not all of the stories our maps tell are about physical conditions or measurable characteristics. Instead, some maps try to change the meanings that people attach to our community, communicating a sense of resources and assets, rather than the sense of danger or problems that the media usually shows.

A participatory approach to using GIS as a negotiating technology is especially powerful because it incorporates information from community residents. We get tremendous input from residents about the real conditions of the area, block by

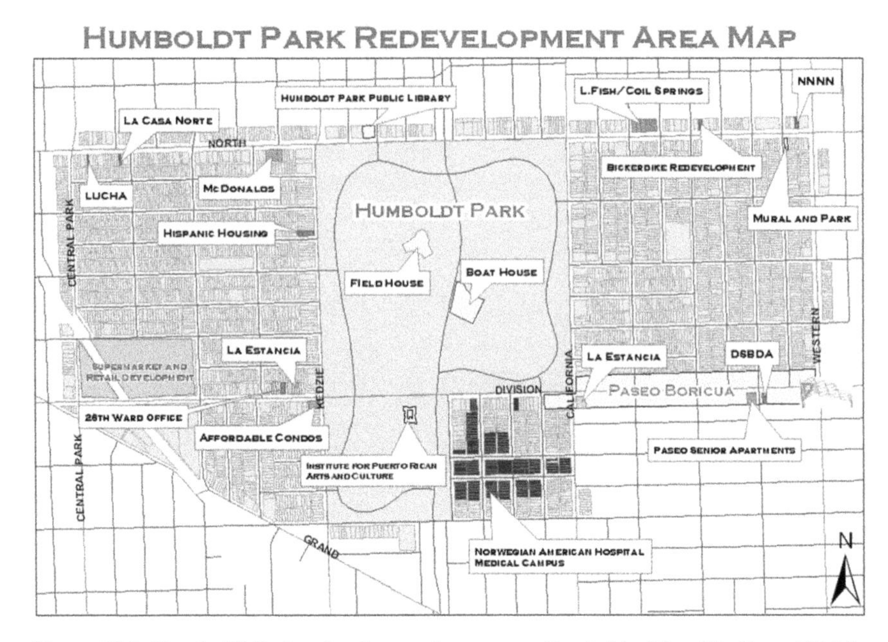


*Figure 20.2* Humboldt Park redevelopment area map. Created by Near Northwest Neighbourhood Network/Humboldt Park Empowerment Partnership

block; a treasure of insight that no one else can give. Including their knowledge makes it easy to produce arguments that defend against those who come in with plans and policies that run counter to community needs and priorities. Approaching PGIS in this way transforms maps and GIS from something that government or universities do to something that community groups and residents can do themselves, as a means to visualise local knowledge and needs or to reinterpret what government data says about the community (see also Sanderson *et al.*, Chapter 15 in this volume).

## Building PGIS partnerships for the long haul: challenges and lessons learned

Some of the challenges we have encountered in the HPCGIS project are common to university-community partnerships, and others are unique to PGIS. Unsurprisingly, our first challenge was negotiating common goals and building strong working relationships. Past experiences with universities can be a problem, and both the NNNN and the Development Council have worked with university partners who have failed to complete projects or be attentive to community needs. Another common struggle has been the competing demands on time (see also Hume-Cook *et al.*, Chapter 19 in this volume). The NNNN and the Development Council have many active projects and people are stretched even without the HPCGIS activities. Because staff usually have two or three different job responsibilities, student interns can be a great help with GIS activities, especially local data collection. Sometimes the availability of student or faculty participants conflicts with community organisation staff members' other work. The community coordinator's site visits help overcome some of the time barriers, even with just a small investment of time at each organisation. Having a regular schedule helps ensure that we have some dedicated time for GIS each week.

Using a technology like GIS in participatory research brings special challenges. It can take months of hard work to develop skills and data, let alone any maps, while everyone waits to see tangible benefits from the project. Participants' different levels of GIS experience and job responsibilities may mean it takes time to agree how to make best use of the GIS for the community. Because PGIS for urban community development is different from other kinds of GIS use, it can be difficult to find appropriate training and skill-building opportunities for new participants. For example, an intensive multi-day course provided by our software vendor focused on local government data and uses of GIS, making it difficult to see how we might use it in our work, or how we might integrate local knowledge into the GIS. Our weekly work sessions with the community coordinator, and winter and summer GIS workshops, were more helpful to fill in knowledge gaps.

Our most significant ongoing challenge is obtaining the data we need for our GIS activities. Most City of Chicago departments will only share printed maps with non-governmental organisations, but we need digital data so we can integrate it with our own information and observations. But collecting local data is time consuming, and incorporating it into the GIS is more difficult than working with existing government data. In PGIS, though, we can rely on different partners to help with these different data challenges. We sometimes rely on university partners to request data on behalf of the project, and students help with local data preparation.

In our experience, overcoming these challenges is possible when university partners remain actively involved over the long term, and when all participants view the project, the GIS, and the spatial data as resources to be embedded in the broader community. Our collaboration in the HPCGIS project has taught us some important lessons about how to sustain these partnerships and ensure that everyone has access to the data, maps, skills and strategic knowledge being developed. Because we all play different roles in the project, in the community and in our professional lives, we have slightly different perspectives on the most important lessons of the HPCGIS project for other PGIS collaborations.

From the perspective of the community organisation staff, PGIS partnerships are likely to be sustainable and effective if they produce immediately-applicable results, and if university partners do not dictate results. One staff participant writes:

> There has to be a clear understanding that the action and the research produced are usable for the community, not just something to publish. If you do research, it has to be action-oriented to be useful, or it will be thrown away and never really used. It has to be something that the community is going to continue to use. The GIS has become part of what we do in our organisation. We are talking about collaborating with West Humboldt Park to gather

information about a nearby vacant site, and GIS is the one of the most important things that will help us prepare a community plan.[3]

The community organisation participants caution against activities and outcomes that are predetermined by university partners because they may re-create previous activities or simply fail to address community needs. Another community participant writes:

> Many universities already have something in mind and they foist it on your organisation. Fortunately, that is not the case with this project. We grew into the GIS and figured out on our own what was important to us. We just had a situation with another university that wanted to do a plan for our Community Land Trust initiative. They wanted to have big community meetings, do some visioning, and have some focus groups, but we did that two years ago, five years ago, and ten years ago.

Finally, the same staff participants note that for community-based organisations to sustain GIS use, they must build strong networks with other organisations. Recently some participants in the HPCGIS project have taken new jobs at other organisations. We view this not as a loss to the project, but as a way of creating a network of GIS knowledge and skills throughout the community. A former community organisation staff member writes:

> People move around but they move around the same table. I am finding other uses for GIS now that I work for another organisation. My role will be helping them. When they say 'We want to map the manufacturers in Logan Square', I can say, 'Let me show you how much more you can do.'

Another staff member adds:

> It is helpful to have our former staff who know about GIS now working in other organisations. They have GIS skills and they are still in the community and involved with HPEP. This expands the GIS, so that other groups can start creating their own GIS hubs or use the resources in our Community Mapping Lab.

From the perspective of the community organisation directors, sustainable and effective PGIS initiatives depend partly on the actions of institutional leaders. One of the directors writes:

> The executive director must work with the staff so they can appreciate the value of GIS to the organisation. He or she cannot just foist GIS on the staff, but must help think about how maps can enhance their work. For a grant application, the director might say, 'Look at this application. Here is what the "funders" want to know, and with the GIS, we have the capability to produce it.' Also the director can help bring the GIS into conversations with the community. Our staff use maps they make to help the community see what is at stake.

The other director writes:

> Creating interest among the staff who use the GIS and maps is crucial. If the staff understand the potential of GIS, then things become much more effective. But he also notes that executive directors need to ensure the accessibility of the GIS activities, writing: If a new technology is not people friendly, it does not benefit us. We are telling people that everything that we are doing here is something that we want you to be able to take advantage of and understand.

Both directors argue that leaders of organisations have a special role to play in sustaining GIS resources and skills. One writes:

> Part of the job of the executive director is recognising the importance of the project, but also finding a way to keep it for the organisation, and not let it go with staff if they leave. We have to document these resources and how to use them, so that they remain with the organisation. To enhance GIS vis-à-vis our organisations, we need to find a way to put it into the community infrastructure.

From the perspective of the community coordinator, it is important to adopt a flexible approach to GIS learning, and to help new GIS users create useful data and maps as soon as possible. She writes:

> Quite often we hear the phrase 'GIS has a steep learning curve', but in my experience, the steepness of the learning curve depends on the type of training. The community participants have varied levels of computer skills and each person understands and processes information differently. I use a self-paced hands-on approach that matches their learning styles, and select training material customised to each person's skill level. The learning exercises use real-world examples taken directly from their communities, so that each time we meet, the staff members can produce maps to use in their current work. They can see meaningful results. Sometimes new GIS users get discouraged, believing that GIS is another complicated software tool. When I teach the community users something new, I break it down into small steps with measurable goals and step-by-step instructions. Our goal is to help them develop the necessary tools to gather and analyse their own data, taking ownership of their GIS projects. Usually, the word 'challenge' is associated with problems. But we use the definition 'blueprint for success.' What others perceive as obstacles are faced by our participants with eagerness. Nothing is too difficult, but some things need more time and effort to achieve. We focus on what works instead of on 'challenges'.

From the perspective of the university researcher, the most important outcomes of this project are the creative ways that the NNNN and the Development Council use GIS to benefit their community and work to expand the impacts of their GIS resources. She writes:

> Both groups work with GIS in unique and powerful ways that are very different from more traditional GIS applications. Participatory GIS shows us how to move beyond the limits of what academics think of as GIS applications. Their negotiating strategy of interpreting a single map to create multiple stories about the neighbourhood is one example, and this strategy is critical to their ability to influence local government policies that affect the Humboldt Park area. Another reason for the broad impact of the HPCGIS project is their strategy of making the GIS resources available to other organisations in the Humboldt Park area and beyond. Maps they have produced appear in countless places: community newspapers, pamphlets for residents, affordable housing programme flyers, and presentations to community development professionals and activists across the US. Local elected officials have begun to request HPCGIS project maps, creating a powerful opportunity for Humboldt Park organisations to insert their own observations into local government plans and decisions. They even use the GIS for getting local youth interested in possible careers in information technology or community development.

In sum, for community organisations a great deal of the power and impact of PGIS stems from the diversity of ways it can serve as a process for community change. GIS is a particular kind of 'expert' technology, but PGIS partnerships need all kinds of expertise. Partners must value the expertise that everyone brings to the table, be respectful, and understand that the community contributes to the partnership. Everyone brings something, and everyone gains extraordinary knowledge. When we place the students and a community person together to work with GIS, they learn from each other. Students learn something new on our streets and in our offices, and take that back to the university. Professors see that when students come out into the community they learn more, but academics need to be out in the community with the students to see this difference. These kinds of relationships are the ones that need to be built upon. Community residents and organisations open extraordinary doors for universities. People begin to trust the universities and can truly see them not as an institution but as a partner.

## Notes

1 We are grateful for support from NSF Grant # BCS-0652141, colleagues at NNNN and the WHPFCDC, and John Baldridge, Grant Garstka, Jennifer Grant, Jennifer Hampton, Allan Kempson, Samuel Pearson, Danny Shields and Man Wang.

2 In this chapter, we use the term 'community' in the same way it is used by staff and residents in the Humboldt Park area, to refer to the people who live, work, play, go to school and do business in our neighbourhood. We do not use it to imply sameness or agreement among all these people.

3 In collaborating on this chapter, several authors requested anonymity for their direct quotations. For consistency, we have adopted this standard for all of our quotations.

# Part III

# Reflection

# 21 Participatory data analysis

*Caitlin Cahill, based on work with the Fed up Honeys*

The best way to understand something is to try to change it

Kurt Lewin (Greenwood and Levin 1998: 19)

## Introduction

Analysis is an iterative and ongoing feature of Participatory Action Research's (PAR) cycle of action and critical reflection. A consideration of the theory and practice of participatory analysis raises important questions that stretch our understanding of validity, reliability, rigour and interpretation. Founded upon an 'epistemology that assumes knowledge is rooted in social relations and most powerful when produced collaboratively through action' (Fine *et al.* 2003: 173), participatory analysis embraces knowledge production as a contested, fraught process. It assumes there is no one singular or universal truth, and instead emphasises the power of an intersectional analysis that takes difference into account (see also Kindon *et al.*, Chapter 2 in this volume).

In this chapter I address the theory and practice of participatory analysis, pointing to some of the possibilities and tensions involved in this approach. I reflect upon my own experience of facilitating a participatory data analysis process as part of the PAR project, 'Makes Me Mad: Stereotypes of Young Urban Women of Color', developed with six young women (aka 'the Fed Up Honeys') from New York City's Lower East Side. The 'Makes Me Mad' project involved a research process dependent upon writing as the principle mode of representation and analysis. PAR involving other methods and techniques, such as those explored throughout this volume, will inevitably raise different technical issues about and for analysis. Here I focus more generally upon the broad principles and the practical and theoretical issues that structure a participatory analysis process.

## Different approaches to participatory analysis

The forms that participatory analysis takes vary, as it is a situated and collectively negotiated process. Its practice *with*, rather than separate *from*, participants reflects

PAR's commitment to collective knowledge production and largely distinguishes it from traditional 'research as usual' analysis processes. In the latter case, analysis is typically understood to be:

- a separate phase of the research process;
- conducted by experts; and
- a formal process characterised by abstract and formal tests of validity and cross-questioning of data.

For example, in the traditional (quantitative) mode of analysis with which most researchers are familiar – the hypothetico-deductive method – analysis is classified as a distinct stage in the research process following theoretical supposition and data collection, and before drawing conclusions. A qualitative analysis process, by contrast, takes an inductive approach, in that theory is developed throughout the research process. In this sense analysis is iterative and emergent, and theory is grounded within the research process; however, analysis is again usually only done by the researcher (rather than by the participants or research collective) (Charmaz 2006).

In PAR, analysis may span the spectrum from more traditional and qualitative modes to approaches which engage in a process of collective negotiation and interpretation (see Kindon *et al.*, Chapter 2 in this volume). For example, participatory research can take place where only one person (usually the academic researcher) does the analysis (see Kothari's critique 2001); or where some, but not all co-researchers are able or willing to conduct analysis; to where findings are 'verified' with a wider group, but then analysed by the facilitators; and, finally, there are cases where a collective research team is fully involved in all stages of the process. In fact, within a single PAR project there may be instances of these different modes of analysis throughout a process, as in the Makes Me Mad project.

Self analysis, leading to a reworking of self-representation, is one of the most critical contributions of a PAR process; however, this process may not be straightforward or easy for researchers or participants (see Box 21.1).

### Box 21.1 Data analysis as 'torture'

RUBY: Data analysis – I hate it. When doing this part of the project the day just seemed drag.

ANNISSA: It's just torture that's all.

CAITLIN: Why is it torture?

ANNISSA: I don't know it reminds me of school.

RUBY: It's like to over analyse –

ANNISSA: I think I like to speak and put things out there and not have to think about why I said them.

Source: Caitlin Cahill

Critically analysing the contradictions in one's everyday life as part of a research process can be risky and/or painful, hitting 'too close to home' (Cahill 2004: 278). In the project with the Fed Up Honeys, it was not easy for them to read through and code their own writings as part of a formal analysis process to identify the ways they had internalised racist and sexist stereotypes and applied those same representations to others. While it could be argued that engaging the Fed Up Honey research team in this way enhanced the validity of the research findings because they were able to confirm or verify the interpretations of the data, the process of coding was especially complex because the researchers were analysing what *they* themselves had written throughout the project. This process involved reading through their journal writings, coding individually their collective writing, discussing the codes and then re-coding collaboratively while paying attention to the contradictions within everyone's writings.

It was difficult to read through one another's writings and come to terms with how our writings sometimes demonstrated the ways we had ourselves accommodated stereotypes. The process engaged us in a suffocatingly close analysis (Cahill 2007a). What's more, it reproduced a normative model of analysis requiring us to engage in a distanced analysis that did not make sense within the context of our otherwise auto-ethnographic project. But while the Fed Up Honeys really hated doing this formal and discreet kind of data analysis, they enjoyed the fluid recursive process of analysis that took place throughout the research project as part of the critical cycles of action and reflection central to PAR. Even if we did not name it explicitly as 'data analysis', data analysis was in fact implicit in the feedback loops that were an ongoing part of our collaborative reflective practice.

The contrast between these differing approaches to analysis was clear to the Fed Up Honeys, who found the formal content analysis procedure forced, frustrating and somewhat torturous. It contradicted the therapeutic and emancipatory practice we were otherwise experiencing (see Cameron, Chapter 24 in this volume). While I had wanted to introduce the research team to this recognised processes of coding and textual analysis as part of the research training process I offered to build capacity, the Fed Up Honeys wondered whether it was really necessary to 'prove' what we had already identified as part of the research process. The research team resented the imposition of a process that felt forced compared to the more integrated way that analysis and reflection were woven into our collaborative process.

We had, in fact, already engaged in a meaningful process of analysis from the beginning of our project through a more organic form of participatory analysis involving all of us. This process can be understood as a collaborative and constructive process of reflection. For example, as a result of comparing their everyday personal experiences of racism and exclusion, the Fed Up Honeys started a process of social theorising that informed the development of the Makes Me Mad project and the desire to educate other young people of colour. In this sense, the participatory analysis process may be a self-reflexive dialectical practice of social (and personal) transformation (Kemmis and McTaggart 2005: 578). Reflexivity here refers to the dynamic process by which new understandings shift our engagement with the world, and how through changing our world, in turn we understand it differently.

Analysis serves to propel the participatory cycle forward at each turn in the research process – pushing a research team to ask new questions, to engage the differences between their diverse perspectives, to develop their collective theoretical framework, and make sense of their interpretations. Analysis is thus crucial to each stage of the research process: problem identification, data collection, data analysis and the presentation of research findings. But this cyclical analytic–reflective process does not move in only one direction, rather it shifts back and forth between the 'stages' of the research process, spiralling out for a wide-angle view on the theoretical and political implications of the interpretations, and then zooming up close to understand how an interpretation may, or may not, resonate with one's own personal experience.

One way we engaged in an ongoing analysis process in a conscious fashion was through the practice of regular journal writing and focused discussions. Writing was a place for research team members to privately reflect upon different topics and provided them with a jumping off point for sharing perspectives in a collective group discussion. In turn, I took reflective notes on our shared discussions, documenting points of agreement and/or disagreement (see Cahill 2007a). We posted these reflective notes on 'our wall' which served as a collective memory of our findings and analysis that we would return to throughout the research process.

## Checking back and checking in: balancing multiple perspectives and positionalities throughout the research process

The critical reflection upon one's personal experience is the starting point for much PAR. A consideration of how we developed our research questions illustrates how participatory analysis is integrated in the research process from the beginning. For the project to be created collectively by everyone involved, we started out with an open-ended, generative analysis process comparing our interpretations of our everyday life experiences. Our 'data' were the research team's everyday life experiences of their community (broadly defined). Of course, this process will look different depending upon the project being undertaken. For example, some participatory research processes are bounded by a particular theme at the beginning, for example, 'school' or 'fear in public space', and this will then be the starting point for a comparative analysis process (see also Cameron, Chapter 24 in this volume). In the Makes Me Mad project, analysis was one of the first tasks of the research process, reflecting an inductive approach.

While the development of our research questions originally started big, with broad sweeping concerns focused on community, we moved quickly to the issue of racial stereotyping, and then winnowed our way down into questions specifying the relationship between disinvestment and representations of young women of colour. This process reflected a participatory grounded theory approach. The conceptualisation of our research project was a negotiated process of trying to make connections between different competing agendas and perspectives on the research team. How could our research project 'hold' the various concerns of our research team? It was critical to ensure that in our quest for consensus we did not erase the differences between us (see also McFarlane and Hansen, Chapter 11 in this volume). Our analysis process became a matter of 'checking back' – do these questions get at what we want to find out about? How do these questions 'sit' together? And, then, taking a long perspective, we considered the big picture that emerged as we spliced together our questions. It was a messy, involved practice that engaged us in debates over the political and theoretical implications of our questions (see also Hume-Cook *et al.*, Chapter 19 in this volume) Eventually we arrived at a two-part research question:

1 What is the relationship between the lack of resources (for example, the disinvestment in public education) and the stereotypes of young urban women of colour? In what ways does stereotyping affect young women's well-being?
2 How do stereotypes inform the way you explain/characterise/understand yourself and others? How does this then negatively affect the community?

Out of our negotiated process we collectively produced a research framework that was all the more powerful because it knitted together our multiple perspectives. In this way, triangulation, the convergence of multiple points of view as an approach to the verification of findings, was built into a participatory analysis process as part of the ongoing negotiation between different perspectives. Here is where one might address the nexus between social construction and lived experiences, and the micro- and macro-levels of the environment (Cahill 2004). The development of our research questions reflected this intersection, drawing connections between young women's intimate identifications and structural processes of racism.

Participatory data analysis also involves an explicit attention to our various positionalities, in the tradition of feminist research, except that here research participants are also involved in this process. In practice what this meant for the Fed Up Honeys was clarifying that we were each contributing a perspective from our own unique embodied experiences (for example, 'for me, I experienced it like this') so as to break down any claim to authority and to consciously articulate our particular standpoint and what experiences contributed to our perspective (Hardstock 1983). Throughout the participatory analysis process we engaged in a process of comparing perspectives, actively listening, contributing and making note of the subtle differences between points of view (Bhavnani 1994). With this in mind, another method of analysis that was part of our process was the informal approach of 'checking in' with each other and clarifying our interpretations –'Is this what you meant when you said that?' and then drawing out the social, political and/or ethical implications of each other's interpretations.

## Tensions in participatory analysis

As an ongoing iterative approach, participatory data analysis is a profound, deep, meaningful process that is at once exciting and challenging as Annissa, a Fed Up Honey researcher reflects ( see Box 21.2).

**Box 21.2 Valuing confrontation in participatory data analysis**

I have never had such an intense conversational experience. It had a great deal to do with the small size of the room that we were in and the amount of hours that we were all together – but it created an environment that forced confrontation. There was nowhere to run from any disagreement. We all had to wrestle with our opinions and with our reactions to other's opinions and yet still find a way to work together, to incorporate all of our ideas, and to create something together that spoke for all of us. The fact that we were all able to do that – to disagree, to respect each other, and create research that spoke for us – I found to be truly inspiring.

Source: Annissa, Fed Up Honey researcher

We grappled with several concerns within our process and I discuss them here as they may be relevant for others. There is, of course, no one way to address these issues as they will take various forms in different PAR projects.

### *Concerning emotion*

As participatory research starts with personal experiences and concerns, emotional engagement is central to the process. Yet how can we capture this as 'data'? How does an analysis process mark, and attend to, not only the shaky or passionate voices, but also the silences that punctuate our group reflections? (See also Pain *et al.*, Chapter 4 in this volume).

### *Concerning contextual validity*

Triangulation is intrinsic to the engagement of difference and multiplicity within participatory analysis and to common conceptions of validity and reliability. However, we need to think through validity not only in terms of a convergence of perspectives, but also in terms of how our analyses are situated in a global, political, social and/or economic context. In other words, how might our analyses consciously triangulate micro- and macro-interpretations and attempt to capture 'how the intimate and global intertwine' (Pratt and Rosner 2006: 15)? That said, a participatory process implicitly involves 'jumping scales' – moving from personal experiences to social theorising. Paying attention to the interdependence of scales in our analyses, and drawing out these connections, harbours the potential to extend the power of our interpretations.

### *Concerning privacy and representation*

As the issue of representation was central to our research, we paid explicit attention to how our research might be interpreted or misinterpreted by various audiences (see also Hume-Cook *et al.*, Chapter 19 in this volume). We asked ourselves what was safe to say aloud and share with the others within and beyond the research team (Fine *et al.* 2000). Participatory research collectives need to think through issues of vulnerability, surveillance and the politics of coding in our analyses. Recognising the privileges associated with privacy, it is critical to consider the stakes involved in opening up the community to intellectual scrutiny (Chataway 2001; Fine *et al.* 2003; Smith 1999). Research collectives need to think through how the research may be positioned not only within the community, but also within the broader socio-political context.

### *Concerning coherence*

Participatory research's emphasis upon group work and consensus as a basis for action may serve to purify knowledge by 'tidying up' people's messy lives and excluding anything that does not fit into the shared story (Kothari 2001). How can a participatory analysis retain and work productively with resistance and dissent (Kesby *et al.*, Chapter 3 in this volume)? We struggled with how to produce a narrative that reflected the dissonance of a participatory approach. Experimentation with different research products and combinations of authors is one strategy to capture and hold multiple voices and perspectives in productive tension (see also Cahill and Torre, Chapter 23 in this volume).

## Concluding remarks

If PAR is to stay true to its commitment to social change, its transformative potential rests upon a constructive engagement with politics and positionality within the research process. With this in mind, I conclude with two critical concepts that I think are relevant to thinking through a socially engaged participatory analysis approach. First, critical psychologist Bhavnani pushes us to theorise not only the strong trends that sweep across data, but also importantly, to interrogate the subtle and significant 'differences' within them as part of a project of 'feminist objectivity' committed to challenging the reinscription of the researched into prevailing notions of powerlessness (Bhavnani 1994: 30; Haraway 1988).

Second, a participatory data analysis process must also take seriously the self-conscious commitment to producing 'counter stories' (Harris *et al.* 2001) challenging the status quo and the hegemonic logic of what is understood as 'natural'. As Harris *et al.* (2001) argue, the counter story disrupts the 'master narratives' which are 'often hard to see until you look under the covers – they are normally labelled as common sense and therefore become invisible in everyday life and academic productions' (2001: 8). Participatory data analysis is thus a critical process of making visible the invisible, as part of a project of producing new subjectivities, knowledge and action (cf. Cameron and Gibson 2005a).

# 22 Participatory learning: opportunities and challenges

*Maria Stuttaford and Chris Coe*

## Introduction

This chapter focuses on the learning component of Participatory Action Research (PAR), and reflects on research undertaken as part of a Sure Start evaluation in the United Kingdom (UK). A fundamental element of the PAR process is that learning takes place for all participants. Learning has traditionally been assumed as an outcome, but in reality is seldom reflected upon by participants.

There are several aspects of the learning component of PAR which are important to acknowledge. First, diverse opportunities for learning exist. In practice, opportunities for 'the researched' to learn are often offered only at the point of data collection and dissemination, rather than throughout the research process. Second, where there are multiple participants, there are multiple knowledges and multiple interfaces for sharing knowledge. Third, there are diverse learning relationships within a PAR event or programme. Fourth, there are diverse motivations for learning. Finally, there are diverse agendas which influence what is done with the learning. We reflect on these aspects of learning in this chapter, using research undertaken with adults and drawing on theories of adult models of learning as expounded by Freire (1996), Kolb (1984) and Knowles (1990).

In writing a reflective chapter, it is important for us to describe the basis for these reflections. We first met as researchers working at a university in the UK and began collaborating because of mutual interests in health research and participatory research methods. While undertaking joint research we were also participants in a diploma course for educators working in the tertiary education sector. During our coffee break discussions we reflected how overt and sensitised we were being about learning with our students, but simply assuming useful learning was taking place with the people with whom we were undertaking participatory research. This chapter has evolved in response to our reflexive approach and reflective practice on learning about and with research participants.

We have not undertaken 'deep' PAR in which the researched have participated equally throughout the research process, from conception of the research question, through data collection, data analysis, dissemination and implementation (see Kesby *et al.* 2005). However, we have used the approach and techniques in a variety of research and evaluation settings in both England (Maria and Chris) and

South Africa (Maria). We use the terms 'researcher' and 'researched' not to endorse a hierarchy of power relations but for descriptive clarity. Simply because we are in an academic context does not mean that we see all 'researchers' as academics. We are all, or should be, equal participants in the research process.

## Learning in Participatory Action Research

Texts on participatory approaches attribute much of the theoretical underpinnings to the writings of Paolo Freire and his work on the consciousness raising, or learning, needed before 'the oppressed' can take action (see Kindon *et al.*, Chapter 2 in this volume). The term *conscientização* 'refers to the learning to perceive social, political and economic contradictions, and to take action against the oppressive elements of reality' (Freire 1996: 17). There are many examples of PAR leading to *conscientização* (see Cameron, Chapter 24 in this volume). However, it can also be argued that the growing popularity of participatory approaches has led to those in powerful positions creating a version of participation that focuses on the production of knowledge, with a loss of emphasis on action and 'consciousness raising' (Brock 2002; Cooke and Kothari 2001a; Kapoor 2005). Although PAR is seen to offer the least opportunity for exploitation as a research approach, it is possible for researchers to use elements of PAR to enhance their own learning, knowledge and skills without fully engaging 'the researched' as co-learners and co-producers of knowledge.

A key criticism of PAR is that it focuses on individuals becoming more self-aware of their knowledge but it is unclear on how this individual conscious raising is transformed into political learning and capability (Williams 2004b). We argue that being explicit about the learning component of PAR is vital to remain true to its emancipatory potential. For the purposes of this chapter, we focus on adult learning defined in terms of experiential learning and andragogy.

Kolb's experiential learning process draws on the work of Freire and is defined as 'the process whereby knowledge is created through the transformation of experience' (Kolb 1984: 38). Kolb (1984) identifies several characteristics of experiential learning as listed in Box 22.1.

Knowles (1990) also draws on Freire to explain the emergence of andragogy in learning theory. The development of the andragogical model is based on the view that the pedagogical model, which has driven childhood education, is inadequate for adult education (Knowles 1990). Andragogy is considered here because the basic premises of participatory research are similar and may provide a means for developing the co-learning element of PAR further, with adults. Knowles (1990) outlines several assumptions about the andragogical model as listed in Box 22.2.

It can be argued that for co-learning to take place, PAR will be most effective where it is used in response to co-learners' needs, rather than the needs of only one party or the other. Taking responsibility for reflection, action and learning is important for co-constructed knowledge in which power is shared.

**Box 22.1 Key characteristics of Kolb's Experiential Learning**

- Learning is a process in which concepts are derived from and modified by experience.
- The learner continuously tests knowledge.
- Learning requires reflection and action in order to resolve the dialectic nature of learning and action.
- Learning involves an holistic adaptation by people to the world.
- Learning involves a transactional relationship between learners and the environment.
- Learning is the process of creating knowledge through the transaction between social knowledge and personal knowledge.

Source: Authors' own analysis

**Box 22.2 Key points of Knowles' Andragogical Theory of Adult Learning**

- Adults need to know why they need to learn something before undertaking to learn it.
- People become adults psychologically when they accept responsibility for their decisions
- Adults have a diversity and depth of experience which means that a group of people undertaking a learning activity need to be recognised as being heterogeneous.
- Adults are often motivated to learn by internal pressures such as increased self-esteem and quality of life.
- Adults are prepared to learn what they need to deal with real-life situations.

Source: Adapted from Knowles 1990

## Reflecting on the learning component of Participatory Action Research with Sure Start

For us to learn using PAR, we adopted a reflexive and reflective approach towards the research and learning process. We endorse Sayer (2000) in believing that knowledge is situated and researchers need to adopt a reflexive stance towards their influence on local knowledge. Furthermore, place influences knowledge and power (Kesby 2005) and there are spaces that promote and/or inhibit participation (Cornwall 2002; 2004a; 2004b; Gaventa 2004; Kesby 2007a). Hervik (1994) argues that reflexivity is a means of transforming social experience in the field into anthropological knowledge – that shared social experience leads to shared social reasoning and the gaining of tacit knowledge about the local environment.

Shared reasoning can be linked to the reflection and action of experiential learning (Kolb 1984) and of 'co-intentional education' (Freire 1996: 51; see also Kindon *et al.*, Chapter 2 in this volume). 'Participants need to engage in critical thinking in which researchers and researched act as "co-investigators"' (Freire 1996: 87). As co-investigators, they explore their themes of reality.

> Knowing will be more valid – richer, deeper, more true to life and more useful – if our knowing is grounded in our experience, expressed through our stories and images, understood through theories which make sense to us, and expressed in worthwhile action in our lives.
>
> Reason 2001: 184

By being critical, the researcher and researched can become more detached from their experiences and look at their knowledge from the other side of boundaries, so as to be able to engage with other knowledge (McGee 2000) expressed through, for example, diagramming and performance (see Alexander *et al.*, Chapter 14, and Cieri and McCauley, Chapter 17 in this volume).

Accepting that PAR requires researchers to be reflective, we now use the insights of Kolb (1984) and Knowles (1990) to reflect on the learning as part of the research with Sure Start. We acknowledge that had this been 'deep' participatory research, the users of Sure Start would also have provided reflections on their learning.

Sure Start is a key element in the Labour Government's initiative designed to tackle health inequalities in the UK. The area based programmes aim to assist children and families by providing education, childcare, health and family support in disadvantaged areas. Evaluation of Sure Start exists both nationally and locally. We were involved in the local evaluation of Sure Start in a multi-cultural Midlands city in England. Although a national initiative, each programme was developed according to local need with an emphasis on parental participation at organisational levels. A participatory research approach was used, which focused on the process and development of the programmes, complementing the outcome driven nature of the national evaluation. As part of the participatory approach, several months were spent at the outset of the evaluation working with service providers and users to identify what they wanted the focus of the evaluation to be. This was an important time for building trust and establishing the ethos of the evaluation.

The Participatory Snowballing with Service Users was one component of the wider local evaluation and used peer interviewing. It had two clear aims: the first was to hear first hand the voices of those deemed to be 'hard to reach' within the Sure Start programmes. The term 'hard to reach' is an over-arching term used in this context to describe those who were eligible to be involved in Sure Start programmes, but for whatever reason, were not currently involved. The term is not ideal, as some groups are possibly not accessed in the right way, rather than being hard to reach; however, it is a term that is both widely used and understood in the UK social policy context.

The second aim of the strategy was to further engage parents in the Sure Start programmes by involving them in overall programme development. Parents who had not previously had an 'enhanced' role within Sure Start were recruited, in order for them to increase their understanding of the organisation as well as their own participation within the programme. The possibilities of being involved were relayed to parents who were then encouraged to volunteer. By engaging parents in this way, we recognised that we had a responsibility to ensure that parents' contributions were maximised and that they received well thought out and responsible support, encouragement and recognition for their time and effort.

Parents attended two training sessions; the first was to enable them to construct a short, semi-structured interview schedule and to prepare them to conduct three interviews in their own community using their prepared interview schedule. At the second session parents were helped to conduct a thematic analysis of their interviews and discussion took place about the use of the results. An outcome we hoped for (although it was not an explicit aim) was that some parents would feel able to share the dissemination and feedback findings to the Sure Start team and possibly, to the organisational and management level of the Sure Start programme, the Partnership Board.

We first reflect on the learning in the Participatory Snowballing with Service Users using Kolb's (1984) characteristics of experiential learning (Box 22.1). It was not only the researchers learning from the service workers, and parents, but also the workers and parents learning from each other. Researchers acknowledged parents as experts and drew heavily on their knowledge, experience and expertise. They also provided parents with additional skills that they 'needed to know' in order to take the next step of hearing, first hand, from non users of Sure Start services. This all took place in a relaxed and informal environment. During this process a discernible shift in confidence took place, which culminated in three parents feeling able to present the results of the research to their respective Partnership Boards.

This process of reflection could be said to be an example of resolving Kolb's dialectic relationship between learning and action and demonstrates a holistic adaptation by people to the world. Parents and researchers remained open to learn about different experiences and contexts; parents shared their individual knowledge with each other, with the researchers, and with interviewees, the community of fellow parents. The following quote illustrates a growing sense of ownership by a parent of the Sure Start programme and how the parent was becoming more active in Sure Start: 'If she [Sure Start non-user] feels lonely she should come out and meet new people, Sure Start is the best place. I invited her to the breast feeding café, she loves it!'

For the researchers the learning was also a process, through the interaction and discussion of the parents imparting their insight and own experiences of getting involved with Sure Start. The subsequent data gathered by these parents illuminated a range of reasons as to why people may choose not to become involved in the programmes. The researchers together reflected on the outcomes and, combining this with other research strategies in the evaluation on the topic of 'hard to reach', gave evidence with which the Sure Start programmes could then develop strategies to tackle this area.

We now reflect on the Participatory Snowballing using Knowles's (1990) andragogical model (Box 22.2). Where relevant, we link it to Kolb's model of experiential learning. Both parents and researchers understood the need to explore further why some people find it hard to access and/or utilise services. The 'need to know' was explored. Devising a strategy to understand the problem was crucial to the success of the initiative, and by consenting to become part of the initiative parents understood the need for skill acquisition in order to accomplish the task in hand. Kolb describes this as the need to establish a transactional relationship between the environment and learning.

According to Knowles, adults will resist situations where the will of others is being imposed on them. As researchers wanting to hear the voices of the hard to reach, we felt it was important to design a method that was open to responding to unheard potential service users. Peer interviewing seemed the most appropriate method for helping to achieve this (see also McFarlane and Hansen, Chapter 11, and Higgins *et al.*, Chapter 13 in this volume). The peer interviewing promoted learning based on experience, as expounded by Kolb. It recognised the heterogeneity and the depth of experience of integrated learning within the parent group as well as providing the hard to reach with the opportunity to contribute to the development of Sure Start. Finally, the participatory process was undertaken in direct response to the real-life situation of parents learning about the services through the peer interview process. In this way the learners could learn and then reflect on whether to act and how to act on their learning.

## Critical theory and the emancipatory potential of learning in Participatory Action Research

As co-researchers and co-learners we strive to take part in a research process that encompasses the following elements to learning:

- diverse opportunities for learning and not just at data collection and the dissemination phases of research;
- multiple participants and their/our multiple knowledge, while not valorising any one knowledge (Mohan 2001);
- multiple interfaces for sharing knowledge;
- diverse learning relationships; and
- diverse motivations for learning.

However, learning may not necessarily lead to action, or it may lead to unsuccessful action; learning may not lead to action in the place it is learnt; action as a result of learning may happen hours, days, months after the learning; learning is situated and is shaped by, for example, pre-existing power relations; and people may need to learn to learn, before learning for action can take place (see Krieg and Roberts, Chapter 18, Hume-Cook *et al.*, Chapter 19, and Elwood *et al.*, Chapter 20 in this volume). How, therefore, do we identify and engage with the learning element of participatory research approaches? We draw on critical theory and reflexivity to consider this question, using examples from our participation in the evaluation of Sure Start described above.

PAR and critical theory may be considered as having similar aims, in that they seek to redress conventional balances of power within the researcher–researched relationship through knowledge creation. Participatory research can be used as 'a vehicle for undertaking applied research, co-constructing a shared knowledge base' (Hill *et al.* 2001: 1). A critical social science is one that is critical of the social practices it studies as well as of other theories, and the guiding principle of critical research is 'an emancipatory interest in knowledge' (Alvesson and Sköldberg 2000: 110).

> Critical realists do not deny the reality of events and discourses; on the contrary, they insist upon them. But they hold that we will only be able to understand – and so change – the social world if we identify the structures at work that generate those events or discourses.
>
> Bhaskar 1986: 2

Furthermore,

> The oppressed, contrary to their oppressors, have a direct material interest in understanding the structural causes of their oppression. The relationship between social knowledge or theory and social ... practice will take the form of an emancipatory spiral in which deeper understanding make possible new forms of practice, leading to enhanced understanding and so on.
>
> Bhaskar 1986: 8

However, the question remains as to how this emancipatory spiral is started, and how learning is started. In the Sure Start evaluation, we do not claim that the process was emancipatory, but we did set out to create an iterative, or 'spiral', learning process which did lead to real, direct changes for some parents, such as those who participated in the Participatory Snowballing (see also Kindon *et al.*, Chapter 2 in this volume).

The Participatory Snowballing exercise illustrates the possible emancipatory nature of participatory research, and subsequent change that may result. During the exercise researchers hoped that parents might feel sufficiently enabled to feed back the results of their research to the Sure Start teams and possibly to the Partnership Board. Whilst discussing dissemination of the research findings, these ideas were suggested and three parents subsequently gave presentations to their respective Partnership Boards. All three were then invited to become parent representatives and sit on the Partnership Board. Parents' involvement in the research led to them becoming actively involved and able to influence local decision making through the Sure Start programmes.

## Conclusions

This chapter has sought to make a contribution to the understanding and practice of PAR by focusing on the learning component. We have begun to consider how we might theorise this learning to maximise the emancipatory potential of PAR and ensure rigorous research with accountable outcomes, including learning.

Reflexivity can be used as a research strategy to address the power inherent in the researcher–researched relationship and as an analytical tool for recognising one's own social standpoints (Haney 2002; Maxey 1999). It is necessary for research to consider the role of reflexive practice in critical research as a way to support co-learning.

In the Sure Start evaluation, the researchers were not researched, but the 'researched' (i.e. the parents) did become researchers through the Participatory Snowballing method. By adopting a critical approach and being reflexive and reflective, it is possible to explicitly engage in learning to enhance the emancipatory potential of research. The different knowledges of parents, service users, service providers and researchers can be negotiated and boundaries collapsed.

Those who have engaged in PAR know that learning takes place and often the evidence of this learning is action. However, many of us are probably able to cite examples of 'PAR' that have not led to sustained consciousness raising and emancipatory action. One way to remedy this is to pay more attention to the characteristics of experiential learning and adult learning as we have mapped them out here. Similarly, many of us may be able to cite successful emancipatory outcomes of PAR. However, perhaps these outcomes could be more effectively achieved if explicit attention is paid to the learning component.

# 23 Beyond the journal article

## Representations, audience, and the presentation of Participatory Action Research

*Caitlin Cahill and María Elena Torre*

### Introduction

Postcolonialist scholars have raised the critical problem of academic research being a conversation of 'us' with 'us' about 'them' (Grande 2004; Kelley 1997; Marker 2006). This process reproduces raced, gendered and classed hierarchies and informs policies which too often reinforce structural and social inequalities. In this way the social and environmental sciences become complicit in producing particular types of subjects. As other chapters in this collection demonstrate, Participatory Action Research's alternative epistemological approach has profound implications for rethinking the politics of representation. Here we tease out some of the complicated questions of representation, audience and the presentation of research (Box 23.1) to explore the slippery relations of power and place inherent in them.

We share experiences from two very different US-based Participatory Action Research (PAR) projects with research collectives primarily composed of people from historically marginalised groups. One project was situated in New York City's gentrifying Lower East Side neighbourhood and the other took place in New York State's Maximum Security prison for women. Through our reflections, we illustrate ways in which PAR can reposition those historically 'studied' as objects

**Box 23.1 Questions about representation, audience and action**

Who has the 'authority' to represent a community's point of view? Who should we speak to? Is there a 'we' within the community being represented, or within the participant research team that can be represented? In what language should research be communicated? What kinds of research products speak to what kinds of audiences? How do we engage new audiences with our research? Should some audiences be privileged? How might the research provoke action? And, further, do the methods and practice of participatory research create enough of a shift in traditional positivist research to 'dismantle the master's house' (Lorde 1984: 112) and contribute to social change?

Source: Caitlin Cahill and Maria Elena Torres

of research as subjects (Freire 1996) through careful consideration of research products and their political effects.

## Speaking back to misrepresentations

The politics of representation are at the heart of the PAR project 'Makes Me Mad: Stereotypes of young urban women of colour'. Developed by six young women (aged 16–22) from New York City's Lower East Side, a.k.a. the Fed Up Honeys, the project is concerned with the relationship between the disinvestment of their community and stereotypical representations of young working-class women (see also Cahill, Chapter 21 in this volume). The 'Makes Me Mad' project offers insights into how a PAR process can create a space for re-negotiating representations, developing strategies for reaching out to new audiences and creating useful research products.

The Makes Me Mad project was inspired by the upsetting discovery of stereotypical profiles of young women in an academic research report produced for funding agencies by a local community-based organisation. This perverse practice is common in non-profit organisations around the world – pathologising the very communities they are serving in order to justify their existence. The report featured a hypothetical profile of a young Latina woman, 'Miranda'. According to the profile, Miranda's future was bleak: a high school drop-out, unemployed single mother with HIV, no job prospects, and caught shoplifting – unless of course, this organisation intervened and 'saved' her from self-destruction. Reading through the pages of the report, the Fed Up Honeys were upset by the blatant conflation and exaggeration of multiple stereotypes in the character of 'Miranda'. Miranda was constructed as a symbol of 'at risk' young women from the Lower East Side 'ghetto', who 'make poor choices', are 'irresponsible' and 'lazy'. Problematically, the discourse of risk emphasises individual choice, yet at the same time erases the structural disparities of race and class.

It was clear that the young women who live in the Lower East Side community were not considered as a potential audience for this report. It was especially difficult for the Fed Up Honeys to accept that this report was associated with a youth service organisation that was supposed to be 'on their side' and working on their behalf. Although the young women were familiar with negotiating stereotypes in their everyday lives, they were not prepared to deal with the report's institutionally sanctioned misrepresentations supported by quantitative data and presented as 'scientific fact.' In an effort to respond to these misrepresentations, the Fed Up Honeys developed a project to 'speak back' (Box 23.2)

In PAR projects, addressing the purpose of research and its intended audiences is ideally a collectively negotiated process. Questions we grappled with included:

- What is the objective of our research?
- Who do we want to speak to?
- Why?

**Box 23.2 Making our voices the centrepiece**

Womyn of color are all the more in need of the space and the encouragement to start shaping their paths within society. Part of the journey starts with womyn of color smashing the skewed pictures of themselves that they see being constantly portrayed and reified in the world that they live in. Participatory Action Research is one such method of making sure that we as womyn of color could control how our voices and our thoughts would be portrayed and interpreted through the lens of research. We crafted a project that made our voices the centerpiece and were able to develop new and innovative approaches to research that are more likely to catch the attention of our peers and urge them to re-think their own self-perceptions and those of their communities.

Source: Fed Up Honeys, cited in Cahill *et al.* 2004: 239

In considering the audiences for our research we quickly moved to a related set of questions:

- Who might benefit from the research?
- Do we as researchers and community members have a right to benefit from it?
- How might this research impact 'our community'?

With these questions in mind, the young women developed a critique of traditional research that suggested it was irrelevant, abstract, or worse, in conflict with their needs.

The group decision to privilege other young women of colour as the most important audience (Box 23.3) had implications for the presentation of our findings. Together, we considered questions around 'products' such as:

- How can we effectively reach out to our audiences?
- What is 'our message'?

We reaffirmed the need to consider the interdependent questions of audience and product together (see also Pain *et al.*, Chapter 4 in this volume). Also, aware that their research would be received by both sympathetic and potentially hostile audiences (see also Hume-Cook *et al.*, Chapter 19 in this volume), the Fed Up Honeys developed a multivalent strategy for communicating their research findings to both 'insiders' and 'outsiders' (to young people, to academics, to policy-makers, to educators, and so on). This strategy included a sticker campaign (see Plate 23.1), a website (www.fed-up-honeys.org), a 'youth-friendly' research report (Rios-Moore *et al.* 2004), a book chapter (Cahill *et al.* 2004), conference presentations, and educational workshops in organisations and schools in the community.

**Box 23.3 The importance of communicating to our peers**

Presumably, the main audience for our research would be people outside of our community because it would be simple to assume that these are the people that are misunderstanding us and are the main consumers of stereotypes of young urban womyn of color. But over the course of our discussions we came to the very difficult realization, that we too were consumers of these negative stereotypes, so we decided that our primary audience should be our peers. If we only communicated with outsiders, that presumes that our peers (and ourselves) don't have the level of agency needed to make change to the predominant perceptions of us, and we strongly disagree with that belief.

Source: Rios-Moore *et al.* 2004: 3

The first product that was developed was also the most provocative: the 'stereotype stickers' (Plate 23.1). The intention was to catch the eye of unsuspecting passers-by in public spaces throughout the Lower East Side and cause them to reflect, or as one researcher, Jasmine, put it, 'to realise what it is we have against us'. Confronting young women of colour and the wider public with this 'in your face' strategy, enabled the researchers 'to rob the discourse of its power ... undermining and destabilising racist stereotypes' (Jackson 2004: 263). The hope was also to upset young women of colour and motivate them – just as we had been inspired to re-think the ways we used, resisted and accommodated stereotypes in our everyday lives.

The process of designing the stickers also forced us to become clear about our purpose and our 'message'. For example, when I (Caitlin) shared a prototype of the sticker with a designer for critical feedback, she suggested that perhaps the skin tone should be made darker so that a viewer would more easily 'read' the image as a young woman of colour. My colleague and I, both white, discussed this as an issue of graphic legibility, without reflecting upon the politics of melanin. Not anticipating the reaction of the sticker designers (members of the Fed Up Honeys), I shared her comments 'innocently' to the group. Box 23.4 shows one researcher's response.

Erica articulated her anger at the suggestion that she was 'not black enough' and that her racial identity as a light-skinned Latina was erased in the binary of black and white. Treating the image in a decontextualised fashion lost sight of the social and political issues of skin colour and the emphasis of the Makes Me Mad project, which was explicitly about paying attention to the diversity, complexity and contradictions of young women of colour.

Here, then, is an example of how PAR creates a critical space for self-representation. Without essentialising 'inside' knowledge as 'pure' or 'the truth', we are reminded, as indigenous researchers have long recognised, that insiders carry knowledge, critiques and a line of vision that is not automatically accessible to outsiders. Young women of colour are less likely to pathologise or romanticise

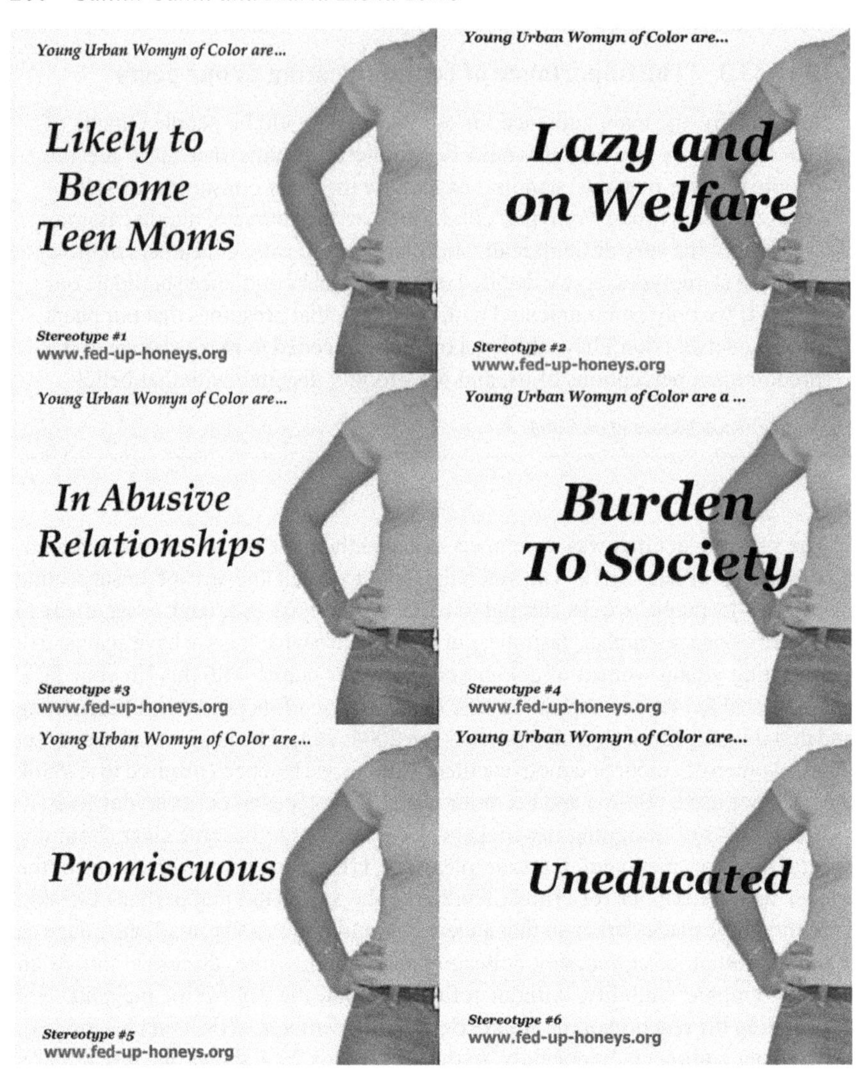


*Plate 23.1* Stereotype stickers (Credit: Fed Up Honeys 2007)

themselves and more likely to understand the ways in which different parts of their lifeworlds are connected (Torre *et al.* 2001). Significantly, a PAR process creates a space and methods for addressing these tensions.

## Representation from behind prison walls

We shift now to a look at questions of representation, audience and the presentation of research in the context of a maximum security prison. In the wake of the 'get tough on crime' policies that proliferated across the US during the 1990s,

**Box 23.4 Dealing with sterotypes**

I never thought I would have to deal with a stereotype like this one …. A woman we approached for honest and critical feedback in the early stages of our research process said she thought that the woman pictured in the (stereotype) stickers did not look dark enough to look like a 'colored' woman. Yes she did. She did say this! And you know that I flipped. I was offended, first of all, because it took a very long and hot seven days of sun in Virginia to get my skin that dark (I am the one on the sticker) … and second of all I AM a womyn of color and for someone to think otherwise is upsetting to me. It wasn't until later that I realized this womyn had just fed into a stereotype …

Source: www.fed-up-honeys.org

hundreds of quietly operating college-in-prison-programmes were shut down. Three years after the demise of one such programme at the Bedford Hills Correctional Facility, inmates organised to begin another new college programme, which would award a degree in sociology.

A number of inmates decided it was necessary to evaluate the new programme to demonstrate its impact and document its value. After much deliberation with prison administrators, it was agreed that a PAR design, while challenging 'behind bars', was necessary (Fine *et al.* 2003). Two graduate students from the Graduate Centre of the City University of New York co-taught a graduate seminar in the prison on Research Methods, reviewing critical research skills. Eight of the students in the course opted to join five Graduate Centre women, to form the College-in-Prison PAR collective. The collective – Kathy Boudin, Iris Bowen, Judith Clark, Aisha Elliot, Michelle Fine, Donna Hylton, Migdalia Martinez, 'Missy', Melissa Rivera, Rosemarie A. Roberts, Pam Smart, María Elena Torre and Debora Upegui – met bi-weekly, over four years. Among us, we represented a rich tapestry of communities: immigrant and US-born; Spanish, English and patois-speaking; lesbian, straight and bisexual; victims and perpetrators of violence; prisoners and 'free' women.

As in the research of the Fed Up Honeys, the College-in-Prison PAR collective was deeply concerned about a population that has suffered a long history of marginalisation, which has included periods of active discrediting and misrepresentation. As a result, the aim of the prison research was two-fold: to document the impact of higher education programmes inside prisons, and to re-present women prisoners to a wider public. For the inmate researchers in the prison, PAR provided opportunities to build new knowledge through research and to present themselves to the world on their terms. It was a welcome, yet complicated project (see Box 23.5).

Whether to protect individuals, groups, programmes or communities, all researchers investigating the consequences of social injustice – such as colonisation or subordination by race, class, gender or sexuality – carefully weigh the implications of which

**Box 23.5 Negotiating self-representation and 'truth'**

An inmate doing research is also a person trying to survive and to get out of prison. This dual reality is always present in the mind of the inmate researcher. As researchers and writers of the research, we are always looking for truths or the closest that we perceive to be 'true.' But for prisoners, important questions include 'Is it safe to say this?' 'What kind of harmful consequences might flow from this either for ourselves personally, or the program or individuals about whom we are writing?' Self-representation is as much a part of being an inmate researcher as 'truth seeking'.

Source: Fine *et al.* 2003: 190

questions are asked, how data is gathered, and how analyses are understood. However, for inmate researchers these ethical choices are further complicated by a tendency for self-censorship that has been described as almost a 'survival instinct' (Fine *et al.* 2003; see also Kesby *et al.*, Chapter 3 in this volume). To begin researching within a prison context we had to contend with a conservative political climate intent on stripping prisons of gym equipment, never mind college programmes. Within this were the daily realities faced by the inmate researchers, who as prisoners lived under constant surveillance where, in the name of 'security,' conversations could be halted, books and notes searched, journals destroyed. As a result, our collective had to grapple with the disproportionate vulnerability experienced by the inmate researchers (see Box 23.6).

As this case demonstrates, participatory researchers need to consider who is made vulnerable through research and how can we protect those communities most impacted by the injustices documented in the research (see also Opondo *et al.*, Chapter 10 in this volume). By explicitly addressing the intended (and unintended) purposes and audiences of research, participatory researchers must grapple with issues of surveillance and privacy. The critical questions of

**Box 23.6 Considering consequences**

As a research collective, we worried that writing something negative about the prison or a program may lead to negative consequences, removing the inmate researchers from a program, from one living unit to another, far from friends or increasing pressure around any of the life-details of living in prison. The inmate researchers were concerned that defining negative truths may create tension between themselves and the women with whom they lived and worked in prison. Their relationships with their peers are a basis for survival as they lived in a closed community in which everything is tied together. There is no exit.

Source: Fine *et al.* 2003: 190

positionality and place are especially significant as members of a participatory research team often represent different communities. While 'inside' and 'outside' may be especially obvious in the case of the prison collective, issues of positionality are relevant for researchers working in tightly knit communities in various contexts around the world such as orthodox religious groups, gangs or remote rural communities.

The thorny question of inside and outside may not always be so starkly delineated but regardless of how it is defined, place matters. One's location – the fluid geographical, political, emotional, gendered, raced, classed position – is, we think, perhaps more important than the distinction between academic and non-academic researcher. Importantly, who is inside and outside is a question that is in flux as we ourselves shift our perspectives and/or standpoints. It is also, importantly, something negotiated collectively. Similarly, participatory researchers make a commitment to negotiate and prioritise the intertwined questions of purpose and audience for their research products (Pain *et al.*, Chapter 4 in this volume).

## Considerations of audience and product

As we moved through data analysis and began thinking about how to present our findings, issues of representation and audience became ever more tangled. Should we lead with our qualitative data, which spoke in intimate detail about what it meant for children to have mothers studying for a Bachelor of Arts degree – often the first person in their families to do so – while in prison? Or would it be more effective for legislators to hear the 'cold hard facts' that any amount of college experience reduces recidivism rates of women in prison from thirty per cent to eight per cent (Fine *et al.* 2001)? The inmate researchers, in particular, were extremely cognisant of the public sentiment regarding crime and prisoners, and cautioned our collective against romanticising inmates or using a highly politicised phrase like 'the prison industrial complex'.

We had very compelling data documenting the profound benefits of college in prison. We wrestled with the questions of how to write our report, whom to ask for endorsements, and where to distribute it. Should our audience be primarily policy-makers? Activists? What about prisoners, college students and faculty? Should the text be written in a single authoritative voice? Or should we create a multi-voiced work filled with the questions and contradictions of participatory work? How would we guard against romanticising women who had been charged with violent crimes? That is, how did we re-present the women with a sense of humanity, re-present their crimes with complexity, and still contextualise the mass incarceration of men and women of poverty and colour in a larger conversation about economic, racial and gendered (in)justice?

Our chief goal was to convince the New York State legislature to restore funds for college in prison programmes. We also wanted to produce research that would be useful for educators, other prisoners and their advocates. In the interest of speaking across audiences, we decided that our primary research product would be a single voiced, multi-method, rigorous and professionally designed report, called Changing Minds, that would also be widely available on a website (www.changingminds.ws).

The report was distributed to every governor in the US and to every New York State Senator and Assembly member. It was also distributed at activist and scholarly meetings on the subjects of prisons, schools, higher education and class, race and gender (in)justice in low-income communities nationally.

The inmate researchers wanted the report to look polished and beautiful. We hired a graphic designer who brought in a draft cover for the report with bold black lettering on a stark white background: CHANGING MINDS. Those of us from the outside loved the drama of the image, but the women inside were disappointed, and argued for a different cover: 'Give it life!', 'Make it sexy; give it lipstick!', 'They already think our life is so drab, make it vibrant'. They were concerned with how they were represented. We all wanted the report to be irresistible, something people would want to touch, hold, place on their coffee tables. The text had to seduce, invite people to read and reacquaint themselves with women inside prison. Moreover, the report had to chip away at the stereotypic images of 'monster women'. The designer returned to the prison a few weeks later, with a brand new version of the report: the cover colourful and strong, the text inside layered with lowered reincarceration rates; cost-benefit analyses; letters; photos; quotes from officers, prisoners and children –- and even postcards. Capturing our desire for the data to jump off the page and move the reader to action, removable postcards were stitched into the report with varied messages like 'Dear Senator – Did you know that college in prison reduces recidivism rates from 30 per cent to 8 per cent? Get tough on crime – Educate Prisoners'. Underneath the postcards were sequenced photographs of women's lives post-release, still photos of lives in motion. The pages surrounding the postcards were draped in quotes from children, 'Now I tell people my mother is away at college!' and corrections officers, 'I'm ambivalent about the college programme, because I can't afford college for myself or my kids. But at least I know there will be less fighting at night, more reading … and the women won't be coming back'.

In addition to our report, we worked closely with activists in the prison reform movement who were interested in materials that would educate broad audiences about college in prison with condensed 'bullet points' and tear-off fundraising sections. We produced 1,000 organising brochures in English and Spanish, carrying the results into communities in a strong voice of advocacy, demanding justice and action. Those of us on the outside have testified in public hearings and worked closely with legislators interested in reopening conversations about prison education. As women from our research team have been released, we have made it a policy to present the work together, yet we still struggle to embody the voice and intellect of those still behind bars.

## Conclusion: beyond the journal article: audience and product

If PAR is to make a meaningful contribution to social change, beyond an 'armchair revolution' (Freire 1996), the impact of our research – action! – is of critical concern. PAR introduces new questions about representation, audience and product that compel us to rethink the role and impact of research. More than an epistemological shift, this approach brings commitments to action that push researchers to work in new and sometimes unfamiliar ways. On a practical level, this means garnering different resources and developing new skills of facilitation and collaboration. At times, this may involve engaging 'experts' to help us carry out our vision; however, the decision of which products to create, and how, may be determined in part by available resources. Participatory researchers with limited financial resources, such as the Fed Up Honeys, may be forced to be creative to develop their own products, and to build upon existing community-based assets to support action agendas. In these projects, a critical component of our work as academics involved developing partnerships and collaborations to create diverse and relevant 'non-academic' products.

As Mitchell and Staeheli (2005: 360) argue, what makes research relevant is shaped not only by those involved in the research but also by the social context in which research is presented, interpreted and used. Understanding the socio-spatial context for participatory practice is critical if we want to position research strategically to make a difference. The impact and receptions of different research products raise important questions for participatory researchers to address in their projects on a case by case basis: what is the research asking readers and audiences to do? What resources are provided to help transform a sense of responsibility into action? Where might the research incite the possibility for change – in the local space where the research was produced or in locations further afield? Frequently, the political activist potential of PAR rests upon its ability to travel 'across' space (see also Kesby *et al.*, Chapter 3 in this volume), forging what Katz (2001: 1231) identifies as a 'grounded but translocal politics', and highlighting the contours of differently situated experiences along which coalitions between diverse places might be built. The challenge for PAR researchers who are serious about social change is to think through how to effectively provoke action by developing research that engages, that reframes social issues theoretically, that nudges those in power, that feeds organising campaigns, and that motivates audiences to change both the way they think and how they act in the world (see also Chatterton *et al.*, Chapter 25 in this volume).

# 24 Linking Participatory Research to Action

## Institutional challenges

*Jenny Cameron*

## Introduction

As the chapters in this book ably illustrate, Participatory Action Research (PAR) is defined by three characteristics: participation, research and action. In this chapter, I explore the action component of PAR and associated institutional challenges, while in the next chapter Paul Chatterton, Duncan Fuller and Paul Routledge examine the challenges of linking action to activism. To explore the institutional challenges I focus on three key questions:

- What actions result from PAR?
- What role do institutions have in the action component of PAR?
- What challenges do institutions present and what strategies can researchers use to negotiate the challenges?

I address these questions by distinguishing between three types of PAR: PAR to liberate and transform research participants; PAR conducted for institutions; and, PAR conducted with institutions (see Figure 24.1).[1] This distinction helps shed light on the actions associated with PAR and the institutional challenges and strategies for managing the challenges. But there are two cautions. The distinction is not intended to set projects against each other, nor to suggest a hierarchy of PAR. Rather, it illustrates the diversity of PAR approaches and how different approaches are appropriate to different situations (see also Kindon *et al.*, Chapter 2 in this volume). Second, some projects have multiple aims and characteristics that fit within two and even three types of PAR. So the three types are not mutually exclusive, but are useful for helping us think about actions, institutional challenges and ways of managing these challenges.

In this chapter, I use examples of PAR projects from across the globe to illustrate the three approaches and highlight the main institutional challenges. Drawing on these examples, and my experiences in Australia, I conclude by outlining strategies for negotiating institutional challenges.

| | *Main characteristics* | *Main challenges* |
|---|---|---|
| PAR type I: liberatory projects. | • Research undertaken with oppressed and exploited groups.<br>• Aims to transform people's day-to-day lives.<br>• Often based on opposing (or exposing) the oppressive and exploitative practices of institutions. | • How do researchers best work with oppressed and exploited groups?<br>• How can findings be used to change institutional practices? |
| PAR type II: researching for institutions. | • Research undertaken on behalf of institutions.<br>• Aims to produce insights and recommendations for institutions to respond to. | • How do researchers retain the liberatory potential of PAR?<br>• How do researchers ensure that institutions act on findings? |
| PAR type III: researching with institutions. | • Representatives from institutions participate as co-researchers.<br>• Aims to build institutional commitment to act on findings | • How do researchers negotiate institutional cultures?<br>• How do researchers ensure that institutions act on findings? |

*Figure 24.1* Three types of PAR: characteristics and challenges for academic researchers (Source: Author 2006)

## Participatory Action Research type I: liberatory projects

The liberatory school of PAR grew out of the popular education movement, especially as practised in the majority world in the 1960s and 1970s (Hall 2005: 6). The approach aims to improve people's circumstances, particularly for those who experience oppressive and exploitative conditions that limit life opportunities. The influence of the popular education movement is evident through the emphasis that is placed on self-transformation as participants learn how their individual experiences of exploitation and oppression are shared by others, and about factors shaping these experiences. Liberatory PAR takes this transformative action a step further by supporting research participants to take collective political action to change their circumstances (see Kindon *et al.*, Chapter 2 in this volume).

In this type of PAR, institutions are frequently seen as the cause of oppression and exploitation, and researchers work with participants to devise ways to challenge or change institutional practices, whether of public bodies (such as national governments or international agencies), private institutions (such as multinational corporations) or cultural institutions (in the form of embedded practices such as racism, sexism and homophobia). Quoss *et al.* (2000) report on a PAR project that challenged a public body, in this case the Wyoming State legislature in the USA. When welfare reform threatened to end the informal practice of welfare recipients attending post-secondary education as an alternative to paid work, students, who were also welfare recipients, participated in a PAR project investigating the benefits of post-secondary education for welfare recipients. The findings were used to lobby the legislature, and as a result new state welfare laws specifically included post-secondary education as a form of work.

In a very different context, residents of the southern Cauca river valley in Colombia challenged the electricity company who seemed to be mischarging for supply (de Roux 1991). Using PAR, the group worked with outside researchers to gather evidence of discrepancies in the company's billing practices, to involve other local residents in devising solutions, and to mobilise people into collective action (such as withholding payment of electricity bills until the company agreed to negotiate). As a result the company acknowledged the electricity problems and established a grievance procedure.

In both these cases, the public and private institutions responsible for the problems being researched were clearly identifiable; however, cultural institutions, such as racism, sexism, xenophobia, homophobia and so on, are less easy to pinpoint. My own PAR projects with colleagues in Queensland and Victoria, Australia have sought to challenge and overturn cultural stereotypes that limit opportunities for those in economically disadvantaged neighbourhoods (Cameron and Gibson 2005a; 2005b). The stereotype of bleak public housing areas populated by single parents and the unemployed, whose lives are on a downward spiral of welfare dependence, poverty and crime (either as perpetrators or victims), was held not only by 'outsiders' but by local policy-makers, programme workers and even residents. We trained residents in techniques to shift their own and their neighbours' focus, and a very different picture emerged, one of people building and contributing to vibrant and creative informal social and economic networks of support. This picture challenged the prevailing stereotype, and provided the basis for community initiatives that built on people's existing activities (see also Chapters 7–12, 15–21, and 23 in this volume).

Confronting institutions to better people's social and economic circumstances is not easy as many of the chapters in this book illustrate. One challenge of the liberatory approach is deciding on the role of the academic researcher. At what point does the academic researcher become involved? To what extent do they drive the project? There are no right answers to these questions. Rather, it will depend on the context. In de Roux's (1991) project, a residents' group was already investigating the electricity problems, and the outside researchers were called upon to assist. As de Roux (1991: 39) notes, the research 'was a procedure for strengthening a process that was already underway, with a dynamic of its own and its own ups and downs'. Whereas in my own projects, academic researchers have 'kicked things off', by approaching key community members, discussing various possibilities, securing commitment, initiating the research and then playing a key facilitator role as the PAR process unfolds.

A second challenge is working within the constraints of the academic institutions. The welfare reform project discussed by Quoss *et al.* (2000) was constrained by college regulations that limited staff and students from lobbying the state legislature. The research team circumvented the regulations by distinguishing between the academic researchers, who facilitated the research, but did not act on research outcomes, and the students/welfare recipients who participated in the research and took action, but as welfare activists, not students. To negotiate this political minefield, the researchers spelt out that '[t]he researcher never engages in direct action; action marks the boundary between researcher and participant' (Quoss *et al.* 2000: 54) (cf Chatterton *et al.*, Chapter 25 in this volume).

These three projects illustrate liberatory-based PAR geared towards directly transforming the lives of those involved. The projects offer different avenues for research practice, from initiating PAR to being invited to support an existing endeavour; from working in a confined researcher role to being involved in all aspects of the research and resulting actions. There is no single correct academic research practice; rather, it is a matter of assessing the context and working in a way that best suits the group and the circumstances, and then managing the associated challenges.

## Participatory Action Research type II: researching for institutions

A seemingly diametrically opposed form of PAR involves researching on behalf of institutions (see also Higgins *et al.*, Chapter 13, and Stuttaford and Coe, Chapter 22 in this volume). This approach is primarily concerned with producing recommendations for institutions to act on, but it may also generate transformations in line with PAR's original liberatory intent. This approach can be seen as a response to the changing academic context in which researchers increasingly rely on contract research funds, and governments pressure researchers to be more accountable for their research activities (Fuller and Kitchin 2004). It also shows how institutions in some places are beginning to accept PAR as a legitimate form of research and to recognise the value of including 'the researched' as co-researchers (although this is far from being widespread, as noted by Manzo and Brightbill, Chapter 5 in this volume).

One challenge for researchers working with PAR in this way is getting institutions to act on findings and recommendations. Sometimes this is fairly straightforward. For example, the Home Office and a local education authority in the UK funded O'Neill *et al.* (2005) to conduct PAR and make recommendations about the educational needs of newly-arrived migrant families. Community members, including young people, participated as co-researchers to assist with questionnaire interviews and make a video featuring young people. Other community members participated in stakeholder meetings where findings and recommendations were discussed. Since the research has finished, the education authority has acted on recommendations, and devised initiatives that include translating documents into the languages of newly-arrived groups, and improving curriculum materials to reflect issues of importance to the African Caribbean and Muslim communities (O'Neill *et al.* 2005: 85-6).

Other projects have found that getting institutions to act on recommendations can be challenging, especially when findings are at odds with the institution's perception of how it operates. In their Home Office-funded PAR project on young homeless people's experience of crime in Newcastle upon Tyne in the UK, Pain and Francis (2003) encountered institutional resistance. The police force in particular found it difficult to accept the finding that homeless young people were often victims rather than perpetrators of crime, and had negative experiences with police (sometimes being subject to police harassment and violence). The police force resisted the idea that they needed training to better understand young people's situations. Pain and Francis (2003: 52) argue that 'success is affected by the willingness of more powerful individuals or organisations to acknowledge the need for change ... especially where it is critical of their own organisational processes and practice.' They also point out that sometimes institutions are not committed to genuine participatory practices, but are more concerned with 'going through the motions', particularly when governments require departments and agencies to consult the public or engage their communities. To address the challenge of securing institutional commitment (especially commitment to act on findings) they recommend 'careful planning from the start' so that institutions are prepared for what may follow.

A second challenge for researchers is achieving the transformative potential of PAR within a research context that is largely set and driven by an institution. Drawing on the project with young homeless people again, Pain and Francis (2003: 51) found it difficult to achieve 'real participation' for a number of reasons. Young homeless people tend to lead transient lives, and so most were only involved for short periods of time in the participatory diagramming technique used in the data collection stage (see Alexander *et al.*, Chapter 14 in this volume). Furthermore, the young people tended to have more pressing concerns than being involved in PAR – such as finding food and shelter. Pain and Francis (2003) conclude that '[d]espite our best efforts we found, like others, that the ideal of participation is seldom achieved, and that fulfilling the key premise of participatory research – effecting change with participants – is fraught with difficulties' (2003:51).

Nevertheless, there are projects where both institutional objectives and transformation of participants has been achieved. In Arizona in the USA, Silver-Pacuilla was funded by a non-profit educational organisation to work with women with disabilities from an adult literacy programme to explore issues related to adult literacy (Silver-Pacuilla and Associates from the Women in Literacy Project 2004). This PAR project resulted in recommendations about how to improve literacy programmes for women with disabilities and a brochure for others entering literacy programmes. The project also transformed the participants. From being labelled (by themselves and others) as disabled and illiterate, the participants were re-positioned as capable researchers. The women presented project findings at public events, 'allowing [them] to be publicly recognised and respected as authors' (Silver-Pacuilla and Associates from the Women in Literacy Project 2004: 45). In the words of one participant, the project 'helped me figure out who I was' (2004: 53).

The success of the Women in Literacy Project raises the issue of how the scale and timeline of PAR projects impacts on outcomes. The literacy project lasted for twelve months, and sixteen women participated in regular discussions and forums.

This type of relatively intimate setting that gives participants time to build relationships may be best suited to achieving institutional goals and transforming research participants. However, researchers who are conducting PAR on behalf of institutions may have to work to tight timeframes, or with large quantities of data. In this case deeper forms of participation may not be practical (see also Kindon *et al.*, Chapter 2 in this volume). Researchers may have to be content with meeting the institution's requirements, putting to one side the aim of directly transforming participants, and instead work closely with the institution to ensure that findings are acted on (and people's lives subsequently transformed).

## Participatory Action Research type III: researching with institutions

In the third form of PAR, members of institutions participate as co-researchers along with other participant groups. This has been one way to address the challenge of getting institutions to act on recommendations, and can generate transformations for those involved and for the institutions themselves.

Wang *et al.* (2004) included policy-makers and community leaders as co-researchers in their Photovoice research to 'provide the political will to support and help implement … recommendations' (2004: 912) (see also Krieg and Roberts, Chapter 18 in this volume). Working in Flint, Michigan in the USA, they involved four research groups: a group of ten young participants from drug programmes, a group of ten young leaders, a group of eleven community activists, and a group of ten local policy-makers and community leaders (a further group of policy-makers and community leaders sat on a guidance committee). Including policy-makers and community leaders as research participants had important spin-offs. They became champions of the project and made sure that results were presented at key forums, including legislative breakfasts, health department meetings and news programmes. Policy-makers started using Photovoice in other projects. The project also brought together people of disparate ages, incomes, backgrounds, experiences and social and geographic networks. The research team anticipates that the longer-term relationships that are likely to develop between these different participants will be the project's most important and transformative outcome (Wang *et al.* 2004).

Unfortunately, such benefits cannot be guaranteed. In my own PAR projects in Australia, working with governmental and non-governmental institutional representatives as co-researchers has posed interesting challenges. In one project considerable time and effort was spent planning the research with elected councillors and bureaucrats, and securing commitment (such as funding and staff input) for the duration of the project and for the small community-run enterprises that would be initiated through the project but need longer-term support (Cameron and Gibson 2005a; 2005b). We thought the relationship with the institution was solid, but when the political landscape changed we lost our champion and the project itself was threatened. We continued, but in a very different climate. As a result the enterprises that were initiated never received the ongoing support that was originally planned for. On reflection we needed to handle the shift in politics differently.

We could have established personal and regular contact between the local residents we employed as co-researchers and the elected councillors. These co-researchers were our best communicators; instead, we relied on impersonal mechanisms like a project newsletter. We could have had more than one institution involved so that when support of one institution diminished we still had another to rely on (we have used this strategy to good effect in subsequent projects).

Being flexible in the face of political ups and downs is critical in this type of PAR. Our project may have had a different outcome had we handled the politics differently. One project that was certainly saved by the academic researchers' ability to adapt was the Community Planning project in East St Louis, USA. Initial efforts of the university to assist economically and socially marginalised neighbourhoods had little impact. Some local leaders were highly critical and felt that academic researchers benefited through grants, publications and promotions while nothing changed in the neighbourhoods. However, one community organisation was interested in working with the university, so long as a more participatory approach was employed. The willingness of academic researchers to work collaboratively and to act on the suggestions of residents has led to a fruitful long-term partnership with neighbourhoods, and governmental and non-governmental institutions in East St Louis (Reardon 1997).

Another challenge for this kind of PAR is the capacity of institutions. At a practical level, staff may have scarce time and resources to be effective co-researchers (despite their desire to be active research partners). This particularly applies with non-government organisations that are often running on a shoestring budget, and have limited staff, or a high staff turnover. Kitchin (2001) faced this challenge when trying to work with disability groups in Ireland and Northern Ireland. Key contacts in two non-governmental organisations left, and remaining staff claimed to know nothing about the research commitments that had been agreed to.

Capacity can also be cultural or ideological. In my area of community-based economic development much is made of building community capacity, but building institutional capacity is also crucial. The PAR work I have done in Australia asks government and non-government institutions to shift from the prevailing focus on needs and deficiencies to a very different approach of identifying, valuing and fostering the skills and strengths that people in marginalised areas already possess: an assets based approach (Cameron and Gibson 2001; Kretzmann and McKnight 1993). Government and non-government institutions are highly attracted to this idea, in principle. In practice, however, staff found it difficult, as their procedures are bound up in a deficit model or needs view of marginalised people and places. They have to switch from managing and leading community development endeavours to supporting local residents to manage and lead. Local residents have been on the receiving end of the deficit approach and readily adopt the assets based approach (and their lives are often transformed as a result). But unless staff in institutions can make the shift they pose the biggest obstacle to achieving tangible outcomes.

A key strategy to address the challenge of building institutional capacity is to spend time working with the institutions involved. In the assets based work that colleagues and I have conducted in Australia, the Philippines and the USA (Cameron and Gibson 2005a; 2005b; Gibson-Graham 2005; 2006) we have found it crucial to set aside time before the research commences to talk about the approach and making sure that staff understand what the project entails. In some cases the time for establishing and consolidating relationships can run into years (see also Pratt *et al.*, Chapter 12, and Hume-Cook *et al.*, Chapter 19 in this volume). But long lead-in times alone do not guarantee success; it is also essential to regularly review progress, acknowledge achievements and discuss difficulties as they arise, particularly when things seem to be going 'off track'. We have also found that it is important for academic staff to spend blocks of time in the field (from two weeks to twelve months) to train and support staff from institutions and community members so they can take a lead role in the research. This sometimes means working out of the offices of institutions so we become part of the day-to-day activities and interactions, an arrangement that has helped us appreciate the context in which the institution operates, and adapt our practices to the institutional setting. When academic staff are not in the field, regular contact is essential using phone, videoconference facilities, email or Internet chat rooms. Such strategies are also appropriate for addressing other challenges of working with governmental and non-governmental institutions (see Figure 24.2).

Academic institutions also present challenges. PAR requires considerable time. Along with the research process itself and the time spent building relationships before the research starts, academic researchers frequently stay connected with research participants beyond the life of the project. The intimacy of PAR means that our lives become bound up with those of our research participants. For example, after being admitted to hospital in the Philippines, one graduate student came to further appreciate the hardships her co-researchers face, and has spent time raising funds in Australia for the hospital (Cahill, pers. comm., 22 December 2006). Academic researchers juggle these pulls on time with the demands of academic life. One key demand is producing academic publications. As Cahill and Torre discuss in Chapter 23, PAR researchers invariably do this academic writing along with developing materials that are relevant to communities and government and non-government institutions. Academic researchers are often highly mobile, particularly if a graduate student or employed on a research grant or contract. This makes it difficult to build the long-term relationships that are often necessary for effective partnerships with institutions and communities. Nevertheless, there are strategies that researchers can use to address these types of academic institutional challenge (see Figure 24.2).

## Conclusion

As my discussion and the contents of this book show, PAR can take different forms, from liberatory-based research, to research for institutions or research with institutions (and other co-researchers). These forms of PAR produce outcomes that range from directly transforming the lives of participants through to transforming the practices of institutions (and indirectly changing people's lives). To bring about

| *Challenges* | *Strategies to manage the challenges* |
|---|---|
| *Governmental and non-governmental institutional challenges* | |
| Getting institutions to act on findings. | • Before the research starts, spend time building commitment to the project and discussing potential outcomes (including negative ones) and institutional responses.<br>• Brief senior members of staff before the research commences.<br>• Become familiar with the institutional context so realistic and appropriate actions are recommended. |
| Meeting expectations of institutions. | • Establish realistic expectations before the research commences.<br>• Regularly document and communicate achievements to the institutions involved (using a format that will grab staff's interest, e.g. oral presentations by community participants).<br>• Use a range of evaluation tools to highlight quantitative and qualitative impacts as the research progresses. |
| Maintaining commitment from institutions involved. | • Before the research commences, agree to a regular meeting schedule to report on progress.<br>• Ensure that senior staff members can attend these meetings.<br>• Use these meetings to "check-in" and make sure that staff understand what is going on, and what is expected of them.<br>• Be prepared to have to deal with difficult issues such the project going "off track".<br>• Regularly communicate achievements (see above). |
| Dealing with changes within institutions (e.g. political changes, staff turnover). | • Work with several institutions so that commitment and skills are built up across a team of research collaborators.<br>• Keep a wider group of institutions informed about the research so they can lend support if necessary. |
| *Academic institutional challenges* | |
| Having to commit large quantities of time to working in the field. | • Train and support local community members and staff from institutions to play a lead role.<br>• Have regular contact with these researchers using phone, videoconference facilities, email or internet chat rooms.<br>• Consider moving to the field for blocks of time. |

*Figure 24.2* Institutional challenges of PAR and strategies for negotiating the challenges (Source: Author 2006)

these outcomes academic researchers manage challenges such as finding appropriate ways of working with research participants, meeting institutional requirements and building institutional commitment to the research process and findings. Perhaps the strongest attribute an academic researcher can have to meet these challenges is commitment and ability to work collaboratively with a range of others.

## Notes

1 In a similar way, Mohan (1999: 43–4) distinguishes between projects that seek to change those involved through empowerment and consciousness raising (a focus on 'the process'), and those that use PAR to extract information for institutions (a focus on 'the product').

# 25 Relating action to activism

## Theoretical and methodological reflections

*Paul Chatterton, Duncan Fuller and Paul Routledge*

### Introduction

As preceding chapters have highlighted, many examples exist of the varying ways in which Participatory Action Researchers work closely with groups to identify needs, plan action research projects, consult and then strive collectively to 'action' their findings. The three of us are all involved to varying degrees in such participatory activities, we would define ourselves not as just as action researchers, but also as 'academic-activists'. Whilst much of our time is spent teaching students, marking essays, and undertaking ever-increasing administrative duties within our increasingly corporatised UK universities, we all consciously strive to bring ourselves into contact with social movement groups struggling for radical social change, and to participate with them in participatory actions and 'research'. More than this, however, we often share the same struggles as resisting others and hence are deeply embedded in social struggles ourselves.

In this chapter we want to reflect upon the similarities and differences between 'our way' of doing things, and how (and why) we think it can offer some important additions to what has come to commonly be regarded as Participatory Action Research (PAR). Some academics have taken the need for what might be termed emancipatory or solidarity research seriously, particularly within our own discipline of geography (for example, Blomley 1994; Maxey 1999). It is through this disciplinary lens that we want to reflect on how we relate action to activism, and what this might mean for the undertaking of PAR more generally within the social and environmental sciences. In a sense, then, our chapter refers to and expands upon the previous discussion about liberatory PAR (see Cameron, Chapter 24 in this volume) and reasserts the importance of solidarity and activism.

### Some beginnings: relevance and radicalism in geography

Discussions about the relevance of the social sciences to 'real-world concerns' have been integral to their development since the second half of the nineteenth century, and the writings of Marx and the anarchist geographer Kropotkin. Such concerns re-emerged in the late 1960s when some academics adopted radical theories and politics rooted in anarchism, Marxism and other critical movements, to facilitate direct involvement by social scientists in the solving of social problems (see for example Berry 1972; Harvey 1972; 1974; Peet 1969; White 1972).

Since the mid- to late 1990s there has been a resurgence of interest in questions of political relevance within many disciplines, alongside a range of other themes focused around exploring the significance of what 'researchers' do. Within geography, there have been powerful critiques of research methodology and the voices or ideas silenced by it, emphasising politically committed research (Nast 1994); increased recognition and negotiation of the differential power relations within the research process (Farrow *et al.* 1995) and multiple activist-academic positionalities (Merrifield 1995); a growing focus on for whom research is produced and whose needs it meets (Nast 1994; Farrow *et al.* 1995); interest in understanding the inter-subjectivity between activist-academics and the researched (McDowell 1992; Staeheli and Lawson 1994; Laurie *et al.* 1999; Moss 2002; Fuller and Kitchin 2004; Chatterton 2006; Cobarrubias 2007); and increasing significance across the social sciences in 'public' and/or 'participatory' variants of sociologies, geographies, anthropologies (for example Burawoy 2004; 2006; the forthcoming special issue of *Antipode* 'Being and Becoming a public scholar'; the People's Geographies project at Syracuse University (www.peoples geographies.org), and the Participatory Geographies Working Group (PyGyWG) of the Royal Geographical Society, UK (www.pygywg.org)).

All of these elements cannot help but make us reflect on doing 'research' in a 'different' way, from its inception, through to dissemination and most crucially (and beyond 'mere' participatory research), its collective action-ing (see Fuller and Kitchin 2004). So as academics we continue to be actively campaigning, making connections, showing solidarity, confronting inequality, supporting local struggles, and seeking progressive social change. But what is the difference between what we do (and are) and those activities undertaken (and embodied) by many Participatory (Action) Researchers? And what, if anything, does this mean for the future significance and practice of PAR? Below are some reflections to continue pushing these debates forward.

## Putting activism back into PAR

For us, activism, academic or otherwise, and PAR are not the same thing, and in this section we want to suggest a few reasons why this might be so and what the implications of such thinking might be for the future of PAR. In so doing we are not necessarily saying that one 'approach' is better than the other, or that the two approaches are unrelated – clearly, activists can be non-participatory while participatory researchers can be very non-active. Both can learn from each other. A key point for us, however, concerns our observation that sometimes, too many times, participatory researchers are more interested in the 'R' than the 'A' in PAR – and probably have a limited repertoire for the 'A' (cf Fals-Borda 2006a).

For us there is a need to reflect on the extent to which those involved in PAR should look beyond 'tools', 'techniques' and 'outputs' and also live up to the challenge of delivering transformative social change (see also Kindon *et al.* Chapter 2 in this volume). Again, this is not to say that PAR is not a diverse tradition where action seeks to address problems faced by marginalised groups and/or disempowered communities. (The chapters in Part II of this volume give a sense of this breadth.) But for us as activists, we wish to raise and revisit questions around the en-action and performance of 'research'. Specifically, we outline below some reflections from our experiences as activists which could be used to explore how PAR can be more than a way of informing policy or improving service delivery, and instead can be used as a vehicle for liberation, radical social transformation and the promotion of solidarity with those defending public services or resisting neoliberal cuts and privatisation.

### *A commitment to social transformation*

The first point we wish to make is one of priorities: why are we doing 'research', and for whom? Have we thought about what 'research' or engagement is actually needed? What or whom is under threat? The overriding motivations of activist 'research' are to develop practice aimed at social transformation rather than to use a set of tools aimed at the 'production of knowledge' and the 'solving' of 'local' problems. Or in the words of Italian Marxist, Conti (2005), 'the goal of research is not the interpretation of the world, but the organisation of transformation' (2005: 2–6).[1]

So, for us (like many of the authors in this volume), 'research', participatory or otherwise, is not just about acquisition, cataloguing, ordering and the publishing of information on groups to help them. It is about jointly producing knowledge with others to produce critical interpretations and readings of the world, which are accessible, understandable to all those involved, and actionable. The differences are clear: significant social transformation does not come through using participatory techniques or appraisal methods to elicit the views of a community facing the construction of a proposed dam by the World Bank, for example. Social transformation requires working with that community to understand how World Bank policy works, so as to resist and possibly reverse the dam's construction.

The possibilities of change often seem daunting, but tools such as scenarios, calendars and campaign planning can help develop energy and enthusiasm (see Trapese 2007). Of course this raises many issues concerning researcher/facilitator neutrality and impartiality, but maybe these issues are not insurmountable for those troubled by them. At the local, session-based scale, good and impartial facilitation remains possible whilst the overall aims of the work remain highly political, strategic and aiming for transformation (see for example, Wadsworth 2000).

### *Beyond participation – developing solidarity*

The second point is: how are we doing our research? The connections and solidarities we have with others involved in resistance are a key part of activist research. Activist research, in our understanding, implies a common identification of problems and desires amongst groups or individuals committed to social change. People may bring different skills and positions to the discussion: some may work in universities, some may not. But the common thread is that there is a desire to work *together* to confront and reverse a set of issues, which have a common effect on all the people concerned. Priorities include jointly identifying exact needs and requirements and how to meet these, using research encounters to promote solidarity and direct forms of democracy not based on hierarchy or domination, and a recognition that objectives may be met through direct action and militancy if needed.

This process may work at different levels of intensity depending on the levels of connection and shared ground. We may already be committed and embedded members of groups and use our research time to support them (see McFarlane and Hansen, Chapter 11 in this volume). We may be approached by groups, which we instinctively support, for help and support. With them we develop a radical critique, which may be used to empower and inform (Pratt *et al.*, Chapter 12 in this volume). Alternatively, we may be approached by groups whose views do not concur with our own political worldviews for specific help to develop ideas and tactics for action. At the end of the day, the key to good activism is finding ways:

- to share relevant and accessible knowledges with groups in ways that don't increase dependency or hierarchy;
- to offer both radical critiques and inspiring alternatives which are translatable and seem doable; and
- to appropriately intervene and criticise, or accept and support.

Putting solidarity into practice also means co-producing contextually relevant knowledges which are useful to groups in their struggles. As several chapters in this volume have demonstrated, these can take the form of plays, artwork, photographs, pamphlets, 'zines', guides or websites, which may be more readily used and understood by the general public.

Solidarity is based on mutual respect and understanding, not agreement for agreement's sake. If real solidarity and mutuality is worked at, respectful critique and disagreement are vital and should develop in a supportive and progressive way that generates solutions beneficial for the group. Disagreements that do arise in terms of both content and tactics can be dealt with using a variety of techniques such as de-escalation, conflict resolution and consensus decision making (see Trapese 2007). It is crucial to evaluate whether disagreements are minor or based on personality differences (and quickly resolvable), or are major and perhaps more structural (and may require reprioritising or group subdivisions). Here, then, we are positing what might be called Solidarity Action Research (SAR) – which is explicit about its aims, its goals, and its desire for strong mutual collaboration.

### *Challenging power relations*

As part of the desire for mutual collaboration, it is important to recognise that presupposing the rigidity of social roles and categories can blind us to the possibilities of common ground, and the potential for transformative dialogue. It is more useful to do away with labels such as 'activist' (which can set certain people up as experts in social change), to allow others to feel they can contribute to social change. Some of the most transformative encounters come through what Giroux (1992) has called 'border pedagogy'. This practice eschews fixed notions of 'us' and 'them' or 'good' and 'bad' tactics. Rather it recognises the many ruptures between groups and embraces and questions differences and newness, however shocking.

As many chapters in this volume demonstrate, challenging power relations means working with groups to uncover structures of power to support people to take control of their own lives. The pedagogical project of Freire (1979; 2004) has been to insist on the dialectical relationship between the 'oppressor' and the 'oppressed'. Through this relationship and the process of *conscientização* we can unpack relationships and causalities, which structure injustice. We recognise our presence in the world, and that history is unfinished business into which we can intervene. Through this process, we can acknowledge that there is not merely external oppression of the 'other', but that we too are subject to oppression and in turn subject others to it. This double movement then compels us to recognise our own role in perpetuating inequality and injustices as well as to tackle larger examples of systematic oppression (Cloke 2002; Kapoor 2005). How this affects our work is crucial and is no easy task, but paying attention to the emotional dimensions of our relationships provides a possible way forward.

### *Building emotional connection*

What Pulido (2003) calls the 'inner life of politics' plays a prominent role in the solidaristic shaping of PAR we are arguing for here (see also Pain *et al*., Chapter 4 in this volume). Transformative encounters based on solidarity often come from our deep emotional responses to the world. These emotions rarely come from academic books or the classroom, but from direct experiences, intuition or a sense of injustice. Newman (2006) talks of the need to inspire defiance and resistance through stories of rebellion and revolt, and then use constructively the frustrations and anger which stem from these. We need to dispel lingering notions that encounters need to be emotionally objective, or that there will be an emotional imbalance between the rational researcher and the emotional subject. Clearly, at some point we need a clear and level head so we can both convey our own emotions and also respect and respond to those of others without losing our cool (see also Cahill, Chapter 21 in this volume). Our emotions can prompt us to ask ourselves about the extent to which we are promoting explicit, cooperative ways of interacting, which are rooted in a deep desire for mutual aid and group support. Also we can ask whether these ways serve to develop a sense of care and responsibility for others, even those we might not even know.

### *Prefigurative action*

The phrase 'be the change you want to see' sums up a prefigurative politics which is common currency amongst activists in the global justice 'movements of movements'. Such a politics rejects blueprints for change in favour of a relational and ethical approach, which accepts that everyone can participate in building change every day. This reformulation reworks the nature of change. It reflects Solnit's (2004: 4) acknowledgement that while most analyses of cause and effect see history as marching forwards, it is more like a crab moving sideways. This sideways movement, or what she calls the 'angel of Alternative History', 'tells us that our acts count, that we are making history all the time' (2004: 75). Our encounters are therefore not just about action in the research process, but how the research process can contribute to wider activism such as protests, demonstrations, events and campaigns to effect change.

Desiring wholesale change often leads to frustrations. We face rhetorical challenges between different ways of organising human life. Brown (2002) suggests that instead of saying, 'You should live in this way!', we need to discuss alternatives by asking, 'Do these alternatives attract you, incite you, make what you've got appear absurd?' She goes on to say: 'You have to incite an interest that has been pounded out of us, an interest in shaping our own lives and the larger orders we live in, you have to incite interest in … freedom' (Brown 2002: 220). This is why a tense encounter or an angry conversation contains hope and has transformative power. There will also be compromises and failures, but these provide resources for thinking through how to enact change in a very complicated, often overwhelming, world.

Rather than offering a future blueprint based on what people 'ought' to be doing, then, the trick is to 'discover tendencies in the present which provide alternative paths out of the current crisis' (Cleaver 1993: 1–16). Some of these will disappear, others will survive, but the challenge remains to find them, encourage people to articulate, expand and connect them: to link and network various micro-politics of resistance.

The ideology of change, therefore, is about movement in which the journey is more important than a hoped-for utopia. Beginning this journey means denouncing how we are living and announcing how we could live (Freire 2004: 105). It means identifying the particular values which will become common currency. This is a tricky process. There are no simple answers, nor should there be. Along the way we will come across ideas and values that might be uncomfortable and unmanageable, but that is the rawness and energy of being involved in social change.

### *Making spaces for action*

Finally, social transformation, unpacking power relations, building solidarity and emotional connections are fine principles, and of course they may all be found in varying extents in P(A)R. However, there have been real limits to their practice due to the lack of spaces within which commonality and connection between disparate groups can be built, and through which change can be en-actioned and distanciated (Kesby 2007a). Finding and extending places for encounter and solidarity in environments unmediated by consumer relations or profit is one of the most significant challenges of our neoliberal times. Oldenberg (1999) talks of the need for 'great good places', which bring people together to dwell and discuss, and similarly Routledge (1996) has introduced the idea of a 'third space for critical engagement', which seeks space beyond the dualism of activists and their others. Chatterton has also suggested that temporary autonomous spaces and social centres have an important role to play here (Chatterton 2002; Chatterton and Hodkinson 2006).

Such spaces provide opportunities for transformative dialogue, mutual learning, as well as conflict. Their openness also makes them sites of potential manipulation, fear and insecurity, but civil society stems from all of these tendencies and should be embraced. It reminds us what it means to be free while also connected. It rejects what stops us from expressing, what restrains us, governs us, disciplines us, and makes us blind to each other and the natural world on which we depend. It is essential that participatory spaces are created for building understanding, encounter and action which are inclusive, which nurture creative interaction with others independent of electoral politics, and which can lead to critical reflection and interventions. How does our work contribute here? Like our colleagues (Kesby 2007a; Kindon and Elwood, forthcoming a), we need to ask ourselves, 'How can we create spaces and conversations that extend past or beyond our research encounters?', and just as crucially, 'How can we open up universities and academic research so they become embedded in the practice of this critical civil society?'

## Conclusions

In this chapter we have been a little brusque, a little passionate, a little confrontational. We hope it made you feel uncomfortable. We have done this as, like most of you, we are passionate about our work and our encounters and we want to use them as tools for transformation. As we suggested at the outset, for us there are too many examples of PAR where it is the 'R' that is all-important – projects and collaborations that go nowhere, have no substantial transformative basis (at any scale), and which do local communities and communities of interest great disservice, if not damage. Some of the ideas we are suggesting will simply not fit into the necessities and parameters of project funding we receive. Clearly, we have to get the balance right between continuing to do research in the social and environmental sciences, whilst also constantly challenging the underlying premises of what we do, why we do it, who it really helps, and whether it is actually needed. Much of this may be out of our control. We have suggested that a commitment to social transformation, challenging power relations, showing solidarity, recognising and using emotions, being the change you want to see, and building spaces for critical dialogue, is crucial to our activist, rather than just action-based, methodology. Confronted as we all are by increasing global social, economic and environmental injustice, this seems to us to be an urgent methodological as well as political imperative.

## Notes

1 This kind of work has a history with the tradition of militant research or workers-inquiry in 1970s Italy where researchers worked closely with workers in the factories to understand the structural origins of the problems and jointly develop solutions. More recently in Argentina the group Colectivo Situaciones has worked closely with the autonomous social movements in Argentina since the 2001 crisis to jointly document, promote and support the autonomous movements (see Colectivo Situaciones 2003; Holdren and Touza 2005).

# Part IV

# Conclusion

# 26 Conclusion: the space(s) and scale(s) of Participatory Action Research

## Constructing empowering geographies?

*Rachel Pain, Mike Kesby and Sara Kindon*

Together with our contributors, writing this book has been an exciting and, at times, demanding project. In it, we set out to discuss ethical, personal and institutional challenges associated with Participatory Action Research (PAR), to link theory and practice through diverse examples from around the world, and to advance the value of a spatial perspective to understanding the connections between people, participation and place.

Throughout, we have tried to make the collection inviting and useful for people with different levels of experience of PAR. At one level, our purpose has been to inform and inspire those who have little experience, and challenge them to consider doing research differently. In each of the book's short chapters we have tried to provide succinct engagements with key theoretical, ethical and political themes that make PAR such an exciting alternative approach to enquiry. At a second level, our task has been to illustrate the wealth and diversity of PAR in ways that will excite even experienced researchers and practitioners to reflect on new issues, try new techniques, and develop and extend their ways of working. At a third level, the book has put forward what we hope are coherent arguments to speak to those who are suspicious or uncertain of PAR, especially in light of recent critiques. We have emphasised our belief that while some forms of participation have been de-radicalised through their institutionalisation and utilitarian application, PAR is still a legitimate and useful *modus operandi* for anyone committed to informing social and environmental transformations through democratic research and action.

There are many ways to read the book. Mechanically, for those not reading from cover to cover but dipping in to review particular debates or techniques, we have tried to provide connections and cross references that draw attention to supporting arguments and linked themes in the book's other sections. Politically, we hope these links will dissuade readers from simply using this book as a source of alternative techniques with which to spice up otherwise untransformed research projects. While we recognise that doing research differently has to begin somewhere – and we have stressed the importance of honesty with participants, our audiences and ourselves – an understanding of the philosophical and political underpinnings of the techniques described in Part II is crucial to their productive deployment. This has been provided in Part I. For readers already making deep participation the core of their research, and those who are more recently taking on the challenge, this book can be read as an incitement to continue to take every opportunity to radically transform research practice and to challenge the institutional structures that privilege dominant ways of knowing and acting (Part III).

One of the ways in which we, as editors, have sought to challenge our own institutional structures has been through our choices about the book's processes of authorship. These have reflected the assertion that PAR is not 'owned' by researchers (particularly not academics), but is shared by any and all who use it. The book, therefore, has been written in part from the perspectives of full-time researchers who are external to marginalised communities, and in part from the perspectives of co-researchers and participants who have deployed PAR to achieve personal, social and environmental change. We suggest that the theoretical, practical and political insights, which they have collectively shared, can enrich the thinking and practice of all our readers, wherever they are sited (Kindon *et al.* Chapter 1). Moreover, we have tried to avoid the impression that empowerment is 'led' by any one set of actors, or that PAR is the only means of empowerment. PAR makes an important, yet modest, contribution to wider movements associated with 'soul power' (Kindon *et al.* Chapter 2), activist resistance (McFarlane and Hansen, Chapter 11; Chatterton *et al.*, Chapter 25) and self-determination (Krieg and Roberts, Chapter 18; Hume-Cook *et al.*, Chapter 19).

## The space(s) of Participatory Action Research

As geographers, we see the potential of, and value in, ideas about space, place and scale now popularised across the social sciences for enhancing the theories and practices associated with PAR, as well as its political impacts. Throughout the book, contributors have highlighted the difference that space, place and scale make to the understandings, practices and outcomes of PAR. Either implicitly or explicitly, they have drawn attention to the materialities of places that affect everyday emotional and sensual encounters; the power-laden nature of spaces which reflect and reshape social, political, research and knowledge regimes; and the connective spaces of dialogue and learning in-between where transformations are located (see especially Kesby *et al.*, Chapter 3; Pain *et al.*, Chapter 4; Kesby and Gwanzura-Ottemoller, Chapter 9; Higgins *et al.*, Chapter 13; Sanderson *et al.*, Chapter 15; Elwood *et al.*, Chapter 20; Cahill, Chapter 21). They remind us that PAR is always already embedded within multiple spaces and scales and it is through our ethical and spatial relationships that we connect people, participation and place to the wider politics of social and environmental transformation. Paying more explicit attention to the spatialities of PAR, we argue, can enhance our theorisations and practice, particularly in relation to its positive and negative power effects and the entangled reach of empowerment.

## The scale(s) of Participatory Action Research

In our view, it is the connectedness and relationality of people, places and processes of participation (Pain *et al.*, Chapter 4) that provide one of the most invigorating aspects of PAR's ability to effect meaningful change and political transformation. Yet all too often our efforts get focused and contained at the scale of the local and the deeper or wider-scale impacts remain un-actioned. While not without its challenges, we would like to propose that our energies now be directed to work at a number of scales, which we conceive of as being flat and intricately connected, rather than hierarchical sites of action (Kesby *et al.*, Chapter 3; Cahill and Torre, Chapter 23).

In some places transformations may be subtle, small-scale, private, personal; in others they may be larger in reach, more spectacular, public. Each may be significant for the lives of people and places involved. And contributors have considered the different changes they may have effected as research partners, from personal and collective transformations of subjectivities, to influencing policy, to altering the codes and practices of repressive institutions.

### *Personal*

PAR can be very satisfying and very challenging. As individuals, we need a range of skills to undertake PAR, many of which are 'soft' and not usually taught formally. Wider life skills drawn from other sites, rather than narrow academic skills, help in implementing and negotiating PAR: including thinking through and around competing ethical dilemmas (Manzo and Brightbill, Chapter 5), sustaining our well-being as well as that of other participants (Adams and Moore, Chapter 6) and, as feminist PAR advocates have underlined, being critically reflexive of our thinking and practice (for example, McFarlane and Hansen, Chapter 11; Sanderson *et al.*, Ch 15; Tolia-Kelly, Chapter 16; Cahill, Chapter 21; Cahill and Torre, Chapter 23).

A key challenge at this scale is to maintain a rigorous critique of the limits of our own praxis, while at the same time trying to deploy participatory approaches in ways that facilitate empowerment and mutual learning. PAR cannot be viewed as a panacea to the power relations endemic in other forms of research; it is a situated mode of knowledge-power with its own limits and power effects (Kesby *et al*, Chapter 3; St. Martin and Hall-Arber, Chapter 7).

### *Inter-personal or relational*

PAR is centrally about relationships, and it succeeds or falls on the quality of these and of the trust that is established. Questions of ethics, researcher positionality, process and reflection can only be answered in relation to (and in dialogue with) the values, positions and experiences of others (see Manzo and Brightbill, Chapter 5; Pratt *et al.*, Chapter 12; Sanderson *et al.*, Chapter 15; Cieri and McCauley, Chapter 17; Hume-Cook *et al.*, Chapter 19; Elwood *et al.*, Chapter 20; Cahill, Chapter 21; Cahill and Torre, Chapter 23).

We need to be alert to the often subtle forms of domination, coercion, inducement and authority that infuse our work even as we struggle to avoid them (see Kesby *et al.*, Chapter 3). In particular, we need to listen to our participants, and note their resistances as barometers of these negative power effects, the better to address them (for example, Alexander *et al.*, Chapter 14; Cahill, Chapter 21), as well as supporting them to take action in their lives as appropriate (for example, Gavin *et al*, Chapter 8; Kesby and Gwanzura-Ottemoller, Chapter 9; Pratt *et al.*, Chapter 12; Sanderson *et al.*, Chapter 15; Krieg and Roberts, Chapter 18; Elwood *et al.*, Chapter 20). These positive power effects legitimate PAR as a practical if partial mode of enquiry (see Kesby e*t al.*, Chapter 3).

Developing work in the arena of these relationalities and emotionalities is critical to understand and further the ethic of care in our work. It is also necessary to extend the effectiveness of our practices to bridge and respect difference, while maintaining diversity and working constructively with resistance and conflict. It is in this domain that we find it helpful to remember the entangled nature of power and empowerment, in both temporal and spatial terms.

### *Institutional*

As many of the chapters in this collection demonstrate, institutions both enable and constrain the changes desired through PAR (Kindon *et al.*, Chapter 2, Gavin *et al.*, Chapter 8; Cameron, Chapter 24 and Chatterton *et al.*, Chapter 25). Within universities (the key type of institution within which many of the contributors to this volume work), despite its increasing popularity and the value of its intellectual heritage and contributions (Kindon *et al.*, Chapter 2; Kesby *et al.*, Chapter 3; Pain *et al.*, Chapter 4), PAR is still marginalised, sidelined as another empirical approach, as 'practice' rather than 'theory' (Kindon *et al.*, Chapter 1) or as 'community service' rather than research. It continues to be misunderstood by institutional ethical review boards (Manzo and Brightbill, Chapter 5; Adams and Moore, Chapter 6), and here action is need from within to change the procedures and measures associated with what constitutes 'ethical research'.

Methodologically, it is also challenging to teach PAR within the confines of university programmes and assessment procedures. While students and lecturers alike may be drawn to PAR's innovative methods to generate rich data, informed and shaped by those who are the subjects of research, there is little guidance or support for those wishing to teach PAR, beyond resource books on particular participatory techniques. There is therefore scope to build on the contributions in this book to develop and to foster a new generation of researchers who pay rigorous attention to the politics and practices of a participatory worldview through the research training they receive.

Beyond universities, we can use PAR to press for a transformation of existing modes of research and institutional arrangements. By 'being the change we want to see' (Chatterton *et al.*, Chapter 25) we can use PAR to challenge the assumptions implicit within a 'Fordist' mode of knowledge production. One way to do this is to continue institutional research activities within and with communities so that 'ordinary people' determine the ethics and safety protocols of research (Manzo and Brightbill, Chapter 5; Adams and Moore, Chapter 6), conduct data collection and analysis (Chapters 7–21), and build theory and disseminate findings (Cahill and Torre, Chapter 23). Another way is to engage and work with organisations and agencies, which have the power to effect change (Gavin *et al.*, Chapter 8; Cameron, Chapter 24). We can do this by scaling 'up' or 'out' our activities and by using the results of PAR to influence trans-institutional and/or trans-governmental agencies as well as local policy-making bodies and campaigning groups. Working in all these ways will help PAR to be deployed and distanciated across a wider institutional–spatial field.

### *Global*

Given the size of the 'translocal' community of Participatory Action Researchers around the world (engaged in connected and collaborative work locally), it is perhaps surprising that they sometimes feel isolated. Many of the challenges identified throughout this book may be context-specific, but many are common across space and can benefit from the advice and support of others. So we should not forget our own needs as a community of researchers dedicated to using PAR (Adams and Moore, Chapter 6) to enhance our theoretical contributions, inform our research practice, learn from our differences, and strengthen our commonalities to be able to effect change beyond our immediate localities.

Further, there remain huge differences in power and privilege between those who deploy PAR in different parts of the world (especially between the minority and majority worlds), and from different sites (communities, policy-making bodies, voluntary organisations, universities and research institutes, and so on). In strengthening our networks of support and solidarity for one another, we must not just think about inclusion of those with fewer resources in our activities, but also learn to transform our own practice by recognising their work and its historical contributions.

With increasingly affordable technologies, the possibilities for greater connection and sharing of our learning and insights are growing. The challenge remains to find ways to harness these technologies so that they work to support PAR's emancipatory potential, bridge cultural and institutional differences and further diverse, grounded and translocal political engagements.

## Constructing empowering geographies?

We began this book with the observation that PAR has recently come in from the cold to become a leading paradigm within the social and environmental sciences. Our aim has been to more firmly establish the conceptual and practical significance of PAR, particularly by focusing on research which takes the relationships between people and place as its central enquiry. Through their varied chapters, contributors have given a flavour of the wealth and diversity of PAR, showing its relevance and value to all sorts of communities, contexts and issues. Together, they emphasise that PAR is not monolithic or fixed. Rather it has its own geographies associated with the influence of particular theoretical and applied traditions and/or the political importance of certain questions and issues in particular places.

What is common throughout, however, is that contributors have identified PAR's huge transformative potential; and yet, as Chatterton *et al.* (Chapter 25) powerfully remind us, too often this potential goes unrealised. Without the 'A', PAR does not live up to the high hopes of those pioneers who first developed it (Kindon *et al.*, Chapter 2). We therefore conclude this book with a call to experienced and new researchers alike, regardless of their institutional location, to consider the geographies of their movements towards empowerment through PAR. By recognising the roles of space and place within PAR, and PAR's embeddedness within multiple scales, we can be more explicit about the cartographies of our engagements and ensure that all three elements – Participation, Action and Research – combine productively to effect positive political change.

# Bibliography


ACME (2007) 'ACME: An International Journal for Critical Geographies'. Available www.acme-journal.org/ (accessed 31 May 2007).

Adams, W. M. and Hulme, D. (2001) 'If community conservation is the answer in Africa, what is the question?', *Oryx*, 35(3): 193–200.

Ahmed, S. (2004) *The Cultural Politics of Emotions*, Edinburgh: Edinburgh University Press.

Alder, C. and Sandor, D. (1990) 'Youth researching youth', *Youth Studies*, 9(4): 38–43.

Allen, J. (2003) *Lost Geographies of Power*, London: Blackwell Publishing.

Alverson, W., Rodriguez, L. and Moskovits, D. (2001) *Peru: Biabo - Cordillera Azul*. Rapid biological inventories: 02, Chicago: The Field Museum.

Alvesson, M. and Sköldberg, K. (2000) *Reflexive Methodology: New Vistas for Qualitative Research*, London: Sage.

Amnesty International (2004) *Stolen Sisters: A human rights response to discrimination and violence against Indigenous women in Canada*. Available www.amnesty.ca/campaings/resources/amr2000304.pdf (accessed 30 June 2005).

Anderson, B. and Tolia-Kelly, D. (2004) 'Matter(s) in social and cultural Geography', *Geoforum*, 35(6): 669–74.

Anderson, P., Carvalho, M. and Tolia-Kelly, D. (2000) 'Intimate distance: fantasy islands and English lakes', *Ecumene*, 8(1): 112–19.

Arnstein, Sherry R. (1969) 'A ladder of citizen participation', *Journal of the American Institute of Planners*, 35(4): 216–24.

Auret, D. and Barrientos, S. (2004) *Participatory Social Auditing: A Practical Guide to Developing a Gender-Sensitive Approach*, IDS Working Paper No. 237, Brighton: Institute of Development Studies. Available www.ntd.co.uk/idsbookshop/details.asp?id=852 (accessed 29 May 2007).

Avis, H. (2002) 'Whose voice is that? Making space for subjectivities in interviews', in L. Bondi *et al.* (eds) *Subjectivities, Knowledges and Feminist Geographies: The Subjects and Ethics of Social Research*, London: Rowman and Littlefield: 191–207.

Barrientos, S., Dolan, C. and Tallontire, A. (2003) 'A gendered value chain approach to codes of conduct in African horticulture', *World Development*, 31(9): 1511–26.

Beauchamp, T. and Childress, J. (1998) *Principles of Biomedical Ethics*, New York: Oxford University Press.

Berger, J. (1972) *Ways of Seeing*, Harmondsworth: Penguin.

Berry, B. (1972) 'More on relevance and policy analysis', *Area* 4: 77–80.

Bery, R. (2003) 'Participatory video that empowers', in S. White (ed.) *Participatory Video: Images that Transform and Empower*, London: Sage: 102–21.

Bhabha, H. (1994) *The Location of Culture*, New York: Routledge.

Bhaskar, R. (1986) *Scientific Realism and Human Emancipation*, London: Verso.

Bhavnani, K. (1994) 'Tracing the Contours: Feminist Research and objectivity', in H. Afshar and M. Maynard (eds) *The Dynamics of 'Race' and Gender: Some Feminist Interventions*, London: Taylor and Francis: 26–40.

Bingley, A. (2003) 'In here and out there: sensations between self and landscape', *Social and Cultural Geography*, 4(3): 329–45.

Blomley, N. (1994) 'Activism and the academy', *Society and Space*, 12: 383–85.

Boal, A., trans. Adrian Jackson (1992) *Games for Actors and Non-actors*, New York: Routledge.

Bourdieu, P. (2004) 'The forms of capital', in S. Ball (ed.) *The Routledge Reader in Sociology of Education*, London, Routledge Falmer: 15–29.

Braden, S. and Mayo, M. (1999) 'Culture, community development and representation', *Community Development Journal*, 34(3): 191–204.

Braden, S., with Thien Huang, T. (1998) *Video for Development: A casebook from Vietnam*, Oxford: Oxfam UK.

Brechin, S., Wilshusen, P., Fortwangler, C. and West, P. (2002) 'Beyond the square wheel: Toward a more comprehensive understanding of biodiversity conservation as social and political process', *Society and Natural Resources*, 15(1): 41–64.

Broad, B. and Saunders, L. (1998) 'Involving young people leaving care as peer researchers in a health care project: a learning experience', *Research, Policy and Planning*, 16(1): 1–8.

Brock, K. (2002) 'Introduction: knowing poverty: critical reflections on participatory research and policy', in K. Brock and R. McGee (eds) *Knowing Poverty: Critical Reflections on Participatory Research and Policy*, London: Earthscan: 1–13.

Brown, I. and Tandon, R. (1983) 'Ideology and political economy in inquiry: action research and participatory research', *Journal of Applied Behavioral Science*, 19(3): 277–94.

Brown, L. and Strega, S. (eds) (2005) *Research as Resistance: Critical, Indigenous and anti-oppressive approaches,* Toronto: Canadian Scholars' Press/ Women's Press.

Brown, W. (2002) 'An interview with Wendy Brown', in J. Schalit (ed.) *The Anti-capitalism Reader: Anti-market Politics in Theory and Practice, Past, Present and Future*, New York: Akashic Press: 208–27.

Browne, K., with L. Bakshi and A. Low (forthcoming) 'Positionalities: it's not about them, it's about us', in S. Smith, R. Pain, S. Marston and J.P Jones III (eds) *Handbook of Social Geography,* London: Sage.

Brydon-Miller, M. (2001) 'Education, research and action: theory and methods of Participatory *Action Research*', in D. Tolman and M. Brydon-Miller (eds) *From Subjects to Subjectivities: A handbook of participatory and interpretive methods*, New York: New York University Press: 76–94.

Brydon-Miller, M., Greenwood, D. and Maguire, P. (2003) 'Why Action Research?', Action Research, 1(9): 1–28.

Brydon-Miller, M., Maguire, P. and McIntyre, A. (eds) (2004) *Travelling Companions: Feminism, Teaching and Action Research*, Westport: Praeger.

Burawoy, M. (2004) 'Public sociologies: contradictions, dilemmas, and possibilities', paper presented at North Carolina Sociological Association Conference on Social Forces, North Carolina, June 2004.

—— (2006) 'A public sociology for human rights', in J. Blau and K. Iyall-Smith (eds) *Public Sociologies Reader,* Lanham: Rowman and Littlefield: 23–50.

Burgess, J. (1996) 'Focusing on fear: the use of focus groups in a project for the Community Forest Unit, Countryside Commission', *Area*, 28(2): 130–5.

Burgess, J., Limb, M. and Harrison, C. (1988a) 'Exploring environmental values through the medium of small groups: 1. theory and practice', *Environment and Planning A*, 20: 309–26.

—— (1988b) 'Exploring environmental values through the medium of small groups: 2. Illustrations of a group at work', *Environment and Planning A*, 20: 457–76.

Butler, J. (1993) *Bodies that Matter: On the Discursive Limits of Sex*, New York: Routledge.

Butler, R. (2001) 'From where I write: the place of positionality in qualitative writing', in M. Limb and C. Dwyer (eds) *Qualitative Methodologies for Geographers: Issues and Debates,* London: Arnold: 264–78.

Cahill, C. (2004) 'Defying gravity: raising consciousness through collective research', *Children's Geographies*, 2(2): 273–86.

—— (2006) Personal communication, 22 December 2006.

—— (2007a) 'Doing research *with* young people: participatory research and the rituals of collective work', *Children's Geographies*, 5(3) 297–312.

—— (2007b) 'The personal is political: developing new subjectivities in a participatory action research process', *Gender, Place, and Culture*, 14(3): 267–92.

Cahill, C., Arenas, E., Contreras, J., Jiang, N., Rios-Moore, I. and Threatts, T. (2004) 'Speaking back: voices of young urban womyn of color using participatory action research to challenge and complicate representations of young women', in A. Harris (ed.) *All About the Girl: Power, Culture and Identity*, New York: Routledge: 231–42.

Cameron, J. and Gibson, K. (2001) *Shifting Focus: Pathways to Community and Economic Development: A Resource Kit*, Latrobe City Council and Monash University, Victoria. Available www.communityeconomics.org/info.html#action (accessed 12 January 2007).

—— (2005a) 'Participatory action research in a poststructuralist vein', *Geoforum,* 36(3): 315–31.

—— (2005b) 'Alternative pathways to community and economic development: the Latrobe Valley Community Partnering Project', *Geographical Research*, 43(3): 274–85.

Cancian, F. (1993) 'Conflicts between activist research and academic success: participatory research and other strategies', *American Sociologist*, 24(1): 92–107.

Castree, N. and Sparke, M. (2000) 'Professional geography and the corporatization of the university: experiences, evaluations, and engagements', *Antipode,* 32(3): 222–9.

Chambers, R. (1994) 'The origins and practice of participatory rural appraisal', *World Development*, 22(7): 953–69.

—— (1997) *Whose Reality Counts: Putting the Last First*, London: Intermediate Technology Publications.

—— (2002) *Participatory Workshops: A sourcebook of 21 sets of ideas and activities*, London: Earthscan.

Chapin, M. and Threlkeld, W. (2001). *Indigenous Landscapes: A Study in Ethnocartography*, Washington DC: Centre for Native Lands.

Charmaz, K. (2006) *Constructing Grounded Theory*, London: Sage.

Chataway, C. (2001) 'Negotiating the observer–observed relationship: participatory action research', in D. Tolman and M. Brydon-Miller (eds) *From Subjects to Subjectivities: A Handbook of Interpretative and Participatory Methods*, New York: New York University Press: 239–55.

Chatterton, P. (2002) '"Squatting is still legal, necessary and free": a brief intervention in the corporate city', *Antipode*, 34(1): 1–7.

—— (2005) 'Can we teach and research our way out of this mess?', paper presented at the Royal Geographical Society/Institute of British Geographers Annual Conference, London, September 2005.

Chatterton, P. (2006) '"Give up activism" and change the world in unknown ways: or, learning to walk with others on uncommon ground', *Antipode*, 38(2): 259–82.

Chatterton, P. and Hodkinson, P. (2006) 'Autonomy in the city: reflections on the UK social centres movement', *City*, 10(3): 125–45.

Chouinard, V. and Grant, A. (1997) 'On not being anywhere near the "project": Revolutionary ways of putting ourselves in the picture', in L. McDowell and J. Sharp (eds) *Space, Gender, Knowledge: Feminist Readings*, London: Arnold: 147–64.

Christensen, P. (2004) 'Children's participation in ethnographic research: issues of power and representation', *Children and Society*, 18(2): 165–76.

Cieri, M. (forthcoming) 'Robbie McCauley's 'Primary Sources': creating routes to an alternative public sphere', *Social and Cultural Geography*.

Clark, C. and Moss, P. (1996) 'Researching *with*: ethical and epistemological implications of doing collaborative, change-orientated research with teachers and students', *Teachers College Record*, 97(4): 518–48.

Clark, M. (2002) 'Oral History: Art and Praxis', in D. Adams and A. Goldbard (eds) *Community, Culture and Globalization*, New York: The Rockefeller Foundation: 87–105.

Clayton, D. (2000a) 'Biopower', in R. Johnston, D. Gregory, G. Pratt and M. Watts (eds) *The Dictionary of Human Geography, 4th Edition*, Oxford: Blackwell: 48.

—— (2000b) 'Governmentability', in R. Johnston, D. Gregory, G. Pratt and M. Watts (eds) *The Dictionary of Human Geography, 4th Edition*, Oxford: Blackwell: 318.

Cleaver, F. (2001) 'Institutions, agencies and the limits of participatory approaches to development', in B. Cooke and U. Kothari (eds) *Participation: The New Tyranny?*, London: Zed: 36–55.

Cleaver, H. (1993) 'Kropotkin, self-valorization and the crisis of Marxism', *Anarchist Studies*, 2(2): 1–34.

Clegg, S. (1989) *Frameworks of Power*, London: Sage.

Cloke, P. (2002) 'Deliver us from evil? Prospects for living ethically and acting politically in human geography', *Progress in Human Geography*, 26(5): 587–604.

Cobarrubias, S. (2007) 'The academy in activism and activism in the academy: collaborative research methodologies and radical geography'. Available www.euromovements.info/html/radical-geography.htm (accessed 17 June 2006).

Colectivo Situaciones (2003) 'Sobre el militante investigador'. Available http://transform.eipcp.net/transversal/0406/colectivosituaciones/es/print (accessed 15 June 2006).

Collins, J. (2003) *Threads: Gender, Labor and Power in the Global Apparel Industry*, Chicago: University of Chicago Press.

Community Economies Collective (2001) 'Imagining and enacting noncapitalist futures', *Socialist Review*, 28(3/4): 93–135.

Conti, A. (2005) 'Metropolitan proletarian research'. Available www.ecn.org/valkohaalarit/english/conti.htm (accessed 12 June 2006).

Cook-Sather, A. (2002) 'Authorizing students' perspectives: toward trust, dialogue and change in education', *Educational Researcher*, 31(4): 3–14.

Cooke, B. and Kothari, U. (eds) (2001a) *Participation: The New Tyranny?*, London: Zed.

—— (2001b) 'The Case for Participation as Tyranny', in B. Cooke and U. Kothari (eds) *Participation: The New Tyranny?*, London: Zed: 1–15.

Cornwall, A. (1992) 'Body mapping in health RRA/PRA', *RRA Notes*, 16: 69–76.

—— (2002) *Making Spaces, Changing Places: Situating Participation in Development*, IDS Working Paper 170, Brighton: Institute of Development Studies. Available www.ntd.co.uk/idsbookshop/detailsasp?id=714 (accessed 29 May 2007).

—— (2004a) 'Introduction: new democratic spaces? The politics and dynamics of institutionalised participation', *IDS Bulletin*, 35(2): 1–10.

—— (2004b) 'Spaces for transformation? Reflections on issues of power and difference in participation in development', in S. Hickey and G. Mohan (eds) *Participation: From Tyranny to Transformation? Exploring New Approaches to Participation in Development*, London: Zed: 75–91.

—— (2006a) 'Historical perspectives on participation in development', *Commonwealth & Comparative Politics*, 44(1): 62–83.

—— (2006b) 'Development's marginalisation of sexuality: report of an IDS workshop', *Gender and Development*, 14(2): 273–89.

Cornwall, A. and Brock, K. (2005) 'What do buzzwords do for development policy? A critical look at "participation", "empowerment" and "poverty reduction"', *Third World Quarterly*, 26(7): 1043–60.

Cornwall, A. and Jewkes, R. (1995) 'What is participatory research?' *Social Science and Medicine*, 41(12): 1667–76.

Cornwall, A. and Pratt, G. (eds) (2003) *Pathways to Participation: Reflections on PRA*, London: ITDG Publishing.

Craig, G., Corden, A. and Thornton, P. (2000) 'Safety in social research', in N. Gilbert (ed.) *Social Research Update*, Guildford: Department of Sociology, University of Surrey, 20. Available http://sru.soc.surrey.ac.uk/SRU29.html (accessed 29 May 2007).

Craig, W., Harris, T. and Weiner, D. (2002) 'Introduction', in W. Craig, T. Harris and D. Weiner (eds) *Community Participation and Geographic Information Systems*, London: Taylor and Francis: 1–2.

Crampton, J. and Krygier, J. (2006) 'An introduction to critical cartography', *ACME*, 4(1): 11–33.

Crang, M. (2003) 'Qualitative methods: touchy, feely, look-see?', *Progress in Human Geography,* 27(4): 494–504.

Crang, M. and Thrift N. (eds) (2000) *Thinking Space*, London: Routledge.

Crawley, H. (1998) 'Living up to the empowerment claim? The potential of PRA', in I. Guijit and M. Shah (eds) *The Myth of Community: Gender Issues in Participatory Development*, London: Intermediate Technology Publications Ltd.: 24–34.

Crocker, S. (2003) 'The Fogo process: participatory communication in a globalizing world', in S. White (ed.) *Participatory Video: Images that Transform and Empower*, London: Sage: 122–44.

Cultural Survival (1995) 'Geomatics: who needs it?' *Cultural Survival Quarterly*, 18(4).

Davison, J. (2004) 'Dilemmas in research: issues of vulnerability and disempowerment for the social worker/researcher', *Journal of Social Work Practice*, 18(3): 379–93.

Davidson, J., Bondi, L. and Smith, M. (eds) (2005) *Emotional Geographies,* Aldershot: Ashgate.

de Roux, G. (1991) 'Together against the computer: PAR and the struggle of Afro-Colombians for public services', in O. Fals-Borda and M. Anisur Rahman (eds) *Action and Knowledge: Breaking the Monopoly with Participatory Action-Research*, New York: The Apex Press, 37–53.

Dearden, P., Bennett, M. and Johnston, J. (2005) 'Trends in global protected area governance, 1992–2002', *Environmental Management*, 36(1): 89–100.

Demeritt, D. (2005) 'The promises of collaborative research', *Environment and Planning A*, 37(12): 2075–82.

Diller, J. (1999) 'Labour dimensions of codes of conduct, social learning and investor initiative', *International Labour Review*, 138(2): 99–129.

Dolan, C., Opondo, M. and Smith, S. (2003) *Gender, Rights and Participation in the Kenya Cut Flower Industry*, NRI Report No. 2768, Chatham: NRI.

Dowler, L. (2001) 'Fieldwork in the trenches: participant observation in a conflict area', in M. Limb and C. Dwyer (eds) *Qualitative Methodologies for Geographers: Issues and Debates*, London: Arnold: 153–64.

Duncan, J. and Ley, D. (1993) *Place/Culture/Representation,* London: Routledge.

Ellis, C. (2007) 'Telling secrets, revealing lies: relational ethics in research with intimate others', *Qualitative Inquiry,* 13(1): 3–29.

Elwood, S. and Leitner, H. (1998) 'GIS and community-based planning: exploring the diversity of neighborhood perspectives and needs', *Cartography and Geographic Information Systems*, 25(2): 77–88.

Ensign, J. (2003) 'Ethical issues in qualitative health research with homeless youths', *Journal of Advanced Nursing*, 43(1): 43–50.

Fals-Borda, O. (2006a) 'Participatory (action) research in social theory: origins and challenges', in P. Reason and H. Bradbury (eds) *Handbook of Action Research*, London: Sage: 27–37.

—— (2006b) 'The north-south convergence: a 30-year first-person assessment of PAR', *Action Research*, 4(3): 351–58.

Farrow, H., Moss, P. and Shaw, B. (1995) 'Symposium on feminist participatory research', *Antipode*, 27(1): 77–101.

Fielding, M. (2004) 'Transformative approaches to student voice: theoretical underpinnings, recalcitrant realities', *British Educational Research Journal*, 30(2): 295–311.

Fincham, B. (2006) 'Back to the "old school": bicycle messengers, employment and ethnography', *Qualitative Research*, 6(2): 187–205.

Fine, M. (forthcoming) 'An epilogue of sorts ...' in J. Cammarota and M. Fine (eds) *Revolutionizing Education: Youth Participatory Action Research in Motion*, London: Routledge.

Fine, M., Weis, L., Weseen, S. and Wong, M. (2000) 'For whom? Qualitative research, representations and social responsibilities', in N. Denzin and Y. Lincoln (eds) *The Handbook of Qualitative Research*, Thousand Oaks: Sage: 107–32.

Fine, M., Torre, M. E, Boudin, K., Bowen, I., Clark, J., Hylton, D., Martinez, M., Roberts, R. M., Smart, P. and Upegui, D. (2001) *Changing Minds: The Impact of College in Maximum Security Prison,* New York: The Graduate Center of the City University of New York.

Fine, M., Torre, M. E, Boudin, K., Bowen, I., Clark, J., Hylton, D., Martinez, M., 'Missy', Rivera, M., Roberts, R., Smart, P. and Upegui, D. (2003) 'Participatory action research: within and beyond bars', in P. Camic, J. Rhodes and L. Yardley (eds), *Qualitative Research in Psychology: Expanding Perspectives in Methodology and Design*, Washington, DC: American Psychological Association: 173–98.

Fisher, P. and Ball, T. (2003) 'Tribal participatory research: mechanisms of a collaborative model', *American Journal of Community Psychology*, 32(3/4): 207–16.

Foucault, M. (1977) *Discipline and Punish: The Birth of the Prison*, London: Allen Lane.

—— (1978) *The History of Sexuality Volume 1*, London: Penguin.

—— (1984) 'Space, knowledge and power: an interview with Michel Foucault', in P. Rabinow (ed.) *The Foucault Reader: An introduction to Foucault's thought*, London: Penguin: 239–56.

—— (1991) 'Governmentality', in G. Burchall, C. Gordon and P. Miller (eds) *The Foucault Effect: Studies in Governmentality*, London: Harvester Wheatcheaf: 87–104.

Fox, N. (2003) 'Practice-based evidence: towards collaborative and transgressive research', *Sociology*, 37(10): 81–102.

France, A. (2004) 'Young people', in S. Fraser, V. Lewis, S. Ding, M. Kellett and C. Robinson (eds) *Doing Research with Children and Young People*, London: Sage: 175–90.

Freire, P. (1972) *Pedagogy of the Oppressed*, Harmondsworth: Penguin.

—— (1979) *Pedagogy of the Oppressed*, London: Penguin.

—— (1988) 'Creating Alternative Research Methods: learning to do it by doing it', in S. Kemmis and R. McTaggart (eds) *The Action Research Reader*, Geelong, Australia: Deakin University Press: 291–313.

—— (1996) *Pedagogy of the Oppressed*, revised edn, London: Penguin Books.

—— (2004) *Pedagogy of Indignation*, London: Paradigm.

French, S. (ed.) (1994) *On Equal Terms: Working with Disabled*, London: Butterworth Heineman.

Friedmann, J. (1992) *Empowerment: The Politics of Alternative Development*, Oxford: Blackwell Publishers.

Frost, N. and Jones, C. (1998) 'Video for recording and training in participatory development', *Development in Practice*, 8(1): 90–4.

Fuller, D. and Kitchin, R. (2004) 'Radical theory/critical praxis: academic geography beyond the academy?', in D. Fuller and R. Kitchin (eds) *Radical Theory/Critical Praxis: Making a Difference Beyond the Academy?*, Vernon and Victoria, BC, Canada: Praxis (e)Press: 1–20.

—— 'Geography and the participatory turn', paper presented at the International Geographical Union, Glasgow, January 2004.

Fuller, D., O'Brien, K. and Hope, R. (2003) *Exploring Solutions to 'Graffiti' in Newcastle upon Tyne*, Newcastle upon Tyne: University of Northumbria.

Garcia, F., Kilgore, J., Rodriguez, P. and Thomas, S. (1995) "It's like having a metal detector at the door': A conversation with students about voice', *Theory into Practice*, 34(2): 138–44.

Gaventa, J. (2004) 'Towards participatory governance: assessing the transformative possibilities', in Hickey, S. and Mohan, G. (eds) *Participation: From Tyranny to Transformation? Exploring New Approaches to Participation in Development*, London: Zed: 25–41.

Gavin, M. (2004) 'Changes in forest use value through ecological succession and their implications for land management in the Peruvian Amazon', *Conservation Biology*, 18(6): 1562–70.

—— (2007) 'Foraging in the Fallows: Hunting patterns across a successional continuum in the Peruvian Amazon', *Biological Conservation*, 134(1): *64–72.*

Gavin, M. and Anderson, G. (2005) 'Testing a rapid quantitative ethnobiological technique: first steps towards developing a critical conservation tool', *Economic Botany*, 59(2): 112–21.

Gibson-Graham, J.K. (1994) "Stuffed if I know!' Reflections on postmodern feminist social research', *Gender Place and Culture*, 1(2), 205–24.

—— (2003) 'An ethics of the local', *Rethinking Marxism*, 15(1): 49–74.

—— (2005) 'Surplus possibilities: postdevelopment and community economies', *Singapore Journal of Tropical Geography*, 26(1): 4–26.

—— (2006) *A Postcapitalist Politics*, Minneapolis: University of Minnesota Press.

Gilligan, C. (1982) *In a Different Voice*, Cambridge: Harvard University Press.

Giroux, H. (1992) *Border Crossings*, London: Routledge.

Gluck, S. and Patai, D. (1991) 'U.S. academics and Third World women: is ethical research possible?', in Gluck, S. and Patai D. (eds) *Women's Worlds: The Feminist Practice of Oral History*, London: Routledge:137–154.

Gordon, G. and Cornwall A. (2005) 'Participation in sexual and reproductive well-being and rights', *PLA Notes*, 50: 73–80.

Grande, S. (2004) *Red Pedagogy: Native American Social and Political Thought*, Lanham, MD: Rowman and Littlefield Publishers, Inc.

Greenwood, D. (2002) 'Action Research: Unfulfilled promises and unmet challenges', *Concepts and Transformation*, 7(2): 117–39.

—— (2004) 'Feminism and action research: is resistance possible? And, if so, why is it necessary?' in M. Brydon-Miller, P. Maguire and A. McIntyre (eds) *Travelling Companions: Feminism, Teaching and Action Research*, Westport: Praeger: 157–68.

Greenwood, D. and Levin, M. (eds) (1998) *Introduction to Action Research: Social Research for Social Change*, Thousand Oaks: Sage.

Gregory, S. (2005) 'Introduction', in S. Gregory, G. Caldwell, R. Avni and T. Harding (eds) *Video for Change: A Guide for Advocacy and Activism*, London: Pluto, Press in association with Witness: xii – xvii.

Gregson, N. (2003) 'Reclaiming 'the social' in social and cultural geography', in K. Anderson, M. Domosh, S. Pile and N. Thrift (eds) *Handbook of Cultural Geography*, London: Sage: 43–57.

Grillo, R. (2002) 'Anthropologists and development', in V. Desai and R. Potter (eds) *The Companion to Development Studies*, London: Arnold: 54–8.

Guidi, P. (2003) 'Guatemalan Mayan Women and participatory visual media', in S. White (ed.) *Participatory Video: Images that Transform and Empower*, London: Sage: 252–70.

Guijt, I. and Shah, M. (1998) 'Waking up to power, conflict and process', in I. Guijt and M. Shah (eds) *The Myth of Community: Gender Issues in Participatory Development*, London: Intermediate Technology Publications: 1–23.

Gwanzura-Ottemoller, F. (2005) ''They tell us we're still young children!' HIV/AIDS related knowledge and the extent and nature of the sexual knowledge and behaviour of Primary School Children in Zimbabwe, unpublished PhD thesis, School of Geography and Geosciences, University of St. Andrews, UK.

Gwanzura-Ottemoller, F. and Kesby, M. (2005) ''Let' talk about sex, baby…': Conversing with Zimbabwean Children about HIV/AIDS.' *Children's Geographies*, 3(4): 201–18.

Hall, B. (1981) 'Participatory Research, Popular Knowledge, and Power: A personal reflection', *Convergence*, 14(3): 6–17.

—— (1997) 'Preface', in S. Smith, D. Williams and N. Johnson (eds) *Nurtured by Knowledge: Learning to do participatory action research*, New York: Apex Press: xiii–xv.

—— (2005) 'In from the cold? Reflections on participatory research from 1970–2005', *Convergence*, 38(1): 5–24.

Haney, L. (2002) 'Negotiating power and expertise in the field', in T. May (ed.) *Qualitative Research in Action*, London: Sage: 286–99.

Hansen, N. (2002) 'Passing through other people's spaces: disabled women, geography and work', unpublished PhD thesis, Department of Geography and Geomatics, University of Glasgow.

Haraway, D. (1988) 'Situated knowledges: the science question in feminism and the privilege of partial perspective', *Feminist Studies*, 14(3): 575–99.

Hardstock, N. (1983) *Money, Sex, and Power*, New York: Longman.

Harris, A., Carney, S. and Fine, M. (2001) 'Counter work: introduction to 'under the covers: theorizing the politics of counter stories'', *International Journal of Critical Psychology*, 4(1): 6–18.

Hart, R. (1992) 'Children's participation: from tokenism to citizenship', *Innocenti Essays No. 4*, Florence: UNICEF.

Hartley, T.W. and Robertson, R. (2006) 'Emergence of multi-stakeholder-driven cooperative research in the Northwest Atlantic: The case of the Northeast Consortium', *Marine Policy*, 30(5): 580–92.

Harvey, D. (1972) 'Revolutionary and counter-revolutionary theory in geography and the problem of ghetto formation', *Antipode*, 4(2): 1–2.

—— (1974) 'What kind of geography for what kind of public policy?', *Transactions of the Institute of British Geographers*, 63(Nov): 18–24.

Hayward, C., Simpson, L. and Wood, L. (2004) 'Still left out in the cold: problematising participatory research and development', *Sociologica Ruralis*, 44(3): 95–108.

Held, V. (ed.) (1995) *Beyond Justice and Care: Essential Readings in Feminist Ethics*, Boulder, Colorado: Westview Press.

Henkel, H. and Stirrat, R. (2001) 'Participation as spiritual duty: empowerment as secular subjection', in B. Cooke and U. Kothari (eds) *Participation: The New Tyranny?*, London: Zed: 168–84.

Herlihy, P. and Knapp, G. (2003) 'Maps of, by and for the peoples of Latin America', *Human Organization*, 62(4): 303–14.

Hervik, P. (1994) 'Shared reasoning in the field: reflexivity beyond the author', in K. Hastrup and P. Hervik (eds) *Social Experience and Anthropological Knowledge*, London: Routledge: 78–100.

Hickey, S. and Mohan, G. (eds) (2004) *Participation: From Tyranny to Transformation? Exploring New Approaches to Participation in Development*, New York: Zed.

—— (2005) 'Relocating participation within a radical politics of development', *Development and Change*, 36(2): 237–62.

Hill, T., Motteux, N., Nel, E. and Paploizou, G. (2001) 'Integrating rural community and expert knowledge through applied participatory rural appraisal in the Kat River Valley, South Africa', *South African Geographical Journal*, 83(1): 1–7.

Hodgson, D. and Schroeder, R. (2002) 'Dilemmas of counter-mapping in community resources in Tanzania', *Development and Change*, 33(1): 79–100.

Holliday, R. (2000) 'We've been framed: visual methodology', *The Sociological Review*, 48(4): 503–22.

Holbrook, B. and Jackson, P. (1996) 'Shopping around: focus group research in North London', *Area*, 28(2): 136–42.

Holdren, N. and Touza, S. (2005) 'Introduction to Colectivo Situaciones', *Ephemera*, 5(4): 595–601.

hooks, b. (1990) *Yearning: Race, Gender and Cultural Politics*, Toronto: Between the Lines.

Hopkins, P. (2006) 'Youth transitions and going to university: the perceptions of students attending a geography summer school access programme', *Area,* 38(3): 240–47.

Horton, B. (1993) 'The Appalachian land ownership study: research and citizen action in Appalachia', in P. Park, M. Brydon-Miller, B. Hall and T. Jackson (eds) *Participatory research in the US and Canada*, Westport: Bergin and Garvey: 85–102.

Howitt, R. and Suchet-Pearson, S. (2006) 'Changing country, telling stories: research ethics, methods and empowerment – working with Aboriginal women', in K. Lahiri-Dutt (ed.) *Fluid Bonds: gender and water, Stree, Kolkata, India*, New York: United Nations Division for Sustainable Development: 48–63.

Hughes, R. (2004) 'Safety in nursing social research', *International Journal of Nursing Studies*, 41(8): 933–40.

Instituto Nacional de Recursos Naturales (INRENA) (2006) *Parque Nacional Cordillera Azul, Plan Maestro (2003–2008)*, Lima-Peru: INRENA.

Irigaray, L. (1985) *This Sex Which Is Not One*, New York: Cornell University Press.

Jackson, J. (2004) 'Racially stuffed shirts and other enemies of mankind: Horace Mann Bond's parody of segregationist psychology in the 1950s', in A. Winston (ed.) *A Measure of Difference: Historical Perspectives on Psychology, Race, and Racism*, Washington DC: American Psychological Association: 261–83.

Jamieson, J. (2000) 'Negotiating danger in fieldwork on crime', in G. Lee-Treweek and S. Linkogle (eds) *Danger in the Field: Risk and Ethics in Social Research*, London: Routledge: 61–71.

Jason, L., Keys, C., Suarez-Balcazar, Y., Taylor, R. and Davis, M. (eds) with Durlak, J. and Hotz Isenberg, D. (2004) *Participatory Community Research: Theories and Methods in Action*, Washington, DC: American Pyschological Association.

Jazeel, T. (2005) ''Nature', Nationhood and the Poetics of Meaning in Rahuna (Yala) National Park, Sri Lanka', *Cultural Geographies*, 12(2): 199–227.

Jensen, O. and Richardson, T. (2003) 'Being on the Map: The New Iconographies of Power over European Space', *International Planning Studies*, 8(1): 9–34.

Johnson, L. and Filemoni-Tafaeono, J. (2003) *Weavings: Women doing theology in Oceania*, Suva: South Pacific Association of Theological Schools.

Jones, E. and SPEECH (2001) ''Of other spaces': Situating participatory practices a case study form South India', IDS Working Paper 137, Brighton: Institute of Development Studies. Available www.ids.ac.uk/ids/bookshop/wp/wp137 (accessed 23 March 2007).

Jupp, E. (2007) 'Participation, local knowledge and empowerment: researching public space with young people', *Environment and Planning A*, 39. (advanced online publication doi:10.1068/a38204 available www.envplan.com/abstract.cgi?id=a38204)

Kapoor, I. (2002) The devil's in the theory: a critical assessment of Robert Chambers' work on participatory development, *Third World Quarterly* 23(1): 101–17.

—— (2005) 'Participatory development, complicity and desire', *Third World Quarterly*, 26(8): 1203–20.

Katz, C. (2001) 'On the grounds of globalization: a topography for feminist political engagement', *Signs: Journal of women in culture and society*, 26(4): 1213–34.

Kelley, R. (1997) *Yo' Mama's Disfunktional! Fighting the culture wars in urban America*, Boston: Beacon Press.

Kelly, D. (1993) 'Secondary power source: high school students as participatory researchers', *The American Sociologist*, 24(1): 8–26.

Kemmis, S. and McTaggart, R. (2005) 'Participatory action research: communicative action and the public sphere', in N. Denzin and Y. Lincoln (eds) *The SAGE Handbook of Qualitative Research*, New York: Sage: 559–604.

Kenya Women Workers Organisation (KEWWO) (2006) 'Labour, employment and social issues in the garment export processing zones in Kenya', Unpublished Report, Nairobi.

Kesby, M. (1999a) 'Beyond the representational impasse? Retheorising power, empowerment and spatiality in PRA praxis', unpublished working paper, University of St. Andrews, UK. Available www.st-andrews.ac.uk/gg/people/kesby/

—— (1999b) 'Locating and Dislocating Gender in Rural Zimbabwe: The making of space and the texturing of bodies', *Gender, Place and Culture*, 6(1): 27–47.

—— (2000a) 'Participatory Diagramming as a means to improve communication about sex in rural Zimbabwe: A pilot study', *Social Science & Medicine*, 50(12): 1723–41.

—— (2000b) 'Participatory Diagramming: Deploying Qualitative Methods through an Action Research Epistemology', *Area*, 32(4): 423–35.

—— (2005) 'Retheorizing Empowerment-through-Participation as a Performance in Space: Beyond Tyranny to Transformation', *Signs: Journal of Women in Culture and Society*, 30(4): 2037–65.

—— (2007a) 'Spatialising Participatory Approaches: The contribution of Geography to a mature debate', *Environment and Planning A*, 39. Advance Online Publication. Available www.envplan.com/epa/fulltext/aforth/a38326 (accessed 26 May 2007)

—— (2007b) 'Editorial: Methodological insights on and from children's geographies', *Children's Geographies*, 5(3) 193–205.

Kesby, M., Fenton K., Boyle, P. and Power, R. (2003) 'An agenda for future research on HIV and sexual behaviour among African migrant communities in the UK', *Social Science and Medicine*, 57(9): 1573–92.

Kesby, M., Kindon, S. and Pain, R. (2005) ''Participatory' Approaches and Diagramming Techniques', in R. Flowerdew and D. Martin (eds) *Methods in Human Geography: A guide for students doing a research project*, London: Pearson Prentice Hall: 144–66.

Kindon, S. (1995a) 'Dynamics of difference: exploring empowerment methodologies with women and men in Bali', *New Zealand Geographer*, 51(2): 10–12.

—— (1995b) 'Gender myths in Bali', in R. Slocum, L. Wichart, D. Rocheleau and B. Thomas-Slayter (eds) *Power, Process and Participation: Tools for Social and Environmental Change*, London: London Intermediate Technology Publications Ltd: 105–9.

—— (1998) 'Of mothers and men: questioning gender and community myths in Bali', in I. Guijt and M. Shah (eds) *The Myth of Community: Gender Issues in Participatory Development*, London: Intermediate Technology Publications: 152–64.

—— (2003) 'Participatory video in geographic research: a feminist practice of looking?', *Area*, 35(2): 142–53.

—— (2005) 'Participatory action research', in I. Hay (ed.) *Qualitative Methods in Human Geography*, Melbourne: Oxford University Press: 207–20.

Kindon, S. and Elwood, S. (forthcoming) 'Widening participation in teaching, learning and research: Participatory action research and Geography teaching', *Symposium Issue of Geography in Higher Education.*

Kindon, S. and Elwood, S. (forthcoming) 'Introduction: More than methods – Reflections or Participatory action research in geographic teaching, learning and research'. *Journal of Geography in Higher Education.*

Kindon, S. and Latham, A. (2002) 'From mitigation to negotiation: ethics and the geographical imagination in Aotearoa/ New Zealand', *New Zealand Geographer*, 58(1): 14–22.

Kindon, S. and Pain, R. (2006) Doing participation geographically. Unpublished paper, available from the authors.

Kinsman, P. (1995) 'Landscape, race and national identity: the photography of Ingrid Pollard', *Area*, 27(1): 300–10.

Kitchin, R. (2000) 'The researched opinions on research: disabled people and disability research', *Disability and Society*, 15(1): 25–47.

—— (2001) 'Using participatory action research approaches in geographical studies of disability: some reflections', *Disability Studies Quarterly*, 21(4): 61–9.

Kitchin, R. and Tate, N. (2000) *Conducting Research in Human Geography: Theory Methodology and Practice*, London: Pearson Prentice Hall.

Kitzinger, J. and Barbour, R. (1999) *Developing Focus Group Research: Policy, Theory and Practice*, London: Sage.

Klodawsky, F. (2007) 'Choosing' participatory research: partnerships in space-time', *Environment and Planning A*, 39. (in press)

Knowles, M. (1990) *The Adult Learner: A Neglected Species*, Houston: Gulf.

Kobayashi, A. (1994) 'Colouring the Field: Gender, 'race', and the politics of fieldwork', *The Professional Geographer*, 46(1): 73–9.

Kolb, D. (1984) *Experiential Learning*, New Jersey: Prentice Hall.

Kothari, U. (2001) 'Power, Knowledge and Social Control in Participatory Development', in B. Cooke and U. Kothari (eds) *Participation: the new tyranny?*, London: Zed: 139–52.

Kothari, U. (2005) 'Authority and expertise: the professionalisation of international development and the ordering of dissent', *Antipode*, 37(3): 402–24.

Kretzmann, J. and McKnight, J. (1993) *Building Communities from the Inside Out: A Path Toward Finding and Mobilizing a Community's Assets*, Evanston, Illinois: The Asset-Based Community Development Institute, Northwestern University. Available http://gearup.ous.edu/gusaccess/documents/pdf/BuildingCommunitiesInsideOut.pdf (accessed 29 May 2007).

Kuokkanen, R. (2004) 'Toward the hospitality of the academy the (im)possible gift of indigenous epistemes', unpublished PhD Thesis, University of British Columbia, Vancouver.

Latham, A. and McCormack, D. (2004) 'Moving cities: rethinking the materialities of urban geographies', *Progress in Human Geography,* 28(6): 701–24.

Laurie, N., Dwyer, C., Holloway, S. and Smith, F. (1999) *Geographies of New Femininities*, Harlow: Longman.

Lawrence, A. (2006) ''No personal motive?' Volunteers, biodiversity, and the false dichotomies of participation', *Ethics, Place and Environment*, 9(3): 279–98.

Lee-Treweek, G. and Linkogle, S. (eds) (2000) *Danger in the Field: Risk and Ethics in Social Research*, London: Routledge.

Lennie, J. (1999) 'Deconstructing gendered power relations in participatory planning: towards an empowering feminist framework of participation and action', *Women's Studies International Forum*, 22(1): 97–112.

Lewin, K. (1946) 'Action research and minority problems', *Journal of Social Issues*, 1–2: 34–6.

Lorde, A. (1984) *Sister Outsider*, Freedom, CA: The Crossing Press.

Lorraine, T. (1999) *Irigaray and Deleuze: Experiments in Visceral Philosophy*, New York: Cornell University Press.

Lunch, N. and Lunch, C. (2006) *Insights into Participatory Video: A Handbook for the Field*, Insight. Available www.insightshare.org/training_book.html (accessed 12 December 2006).

Lykes, M. (2001a) 'Creative arts and photography in participatory action research in Guatemala', in P. Reason and H. Bradbury (eds) *Handbook of Action Research: Participative Inquiry and Practice*, Thousand Oaks: Sage: 1–14.

—— (2001b) 'Activist participatory research and the arts with Maya women: interculturality and situated meaning making', in D. Tolman and M. Brydon-Miller (eds) *From Subjects to Subjectivities: A handbook of participatory and interpretive methods*, New York: New York University Press: 183–99.

McDowell, L. (1992) 'Doing gender: feminism, feminists and research methods in human geography', *Transactions of the Institute of British Geography*, 17(4): 399–415.

McFarlane, H. (2005) 'Disabled women and socio-spatial barriers to motherhood', unpublished PhD thesis, Department of Geography and Earth Sciences, University of Glasgow.

McGee, R. (2000) 'The self in participatory poverty research', in K. Brock and R. McGee (eds) *Knowing Poverty: Critical Reflections on Participatory Research and Policy*, London: Earthscan: 14–43.

McIntyre, A. (2003) 'Through the eyes of women: photovoice and participatory research as tools for reimagining place', *Gender, Place and Culture,* 10(1): 47–66.

McTaggart, R. (ed.) (1997) *Participatory Action Research: International Contexts and Consequences*, New York: State of New York University Press.

Maguire, P. (1987) *Doing Participatory Research: A feminist approach*, Amherst, MA: Centre for International Education, University of Massachusetts.

—— (2000) 'Uneven ground: Feminisms and Action Research', in P. Reason and H. Bradbury (eds) *Handbook of Action Research: Participative inquiry and practice,* Thousand Oaks: Sage: 59–69.

Maguire, P., Brydon-Miller, M. and McIntyre, A. (2004) 'Introduction', in M. Brydon-Miller, P. Maguire and A. McIntyre (eds) *Travelling Companions: Feminism, Teaching, and Action Research*, Westport: Praeger: ix–xix.

Marcuse, P. (1976) Professional ethics and beyond: values in planning, *Journal of the American Institute of Planners*, 42(3): 264–74.

Marker, M. (2006) 'After the Makah whalehunt: indigenous knowledge and limits to multicultural discourse', *Urban Education,* 41(5): 482–505.

Marston, S., Jones, J. and Woodward, K. (2005) 'Human geography without scale', *Transactions of the Institute of British Geographers, New Series*, 30(4): 416–32.

Massey, D. (2005) *For Space*, London: Sage.

Maxey, I. (1999) 'Beyond boundaries? Activism, academia, reflexivity and research', *Area*, 31(3): 199–208.

May, L. (1996) *The Socially Responsive Self: Social Theory and Professional Ethics*, Chicago, Illinois: the University of Chicago Press.

Mayer, V. (2000) 'Capturing cultural identity/creating community: A grassroots video project in San Antonio, Texas', *International Journal of Cultural Studies*, 3(1): 57–78.

Merrifield, A. (1995) 'Situated knowledge through exploration: reflections on Bunge's 'Geographical Expeditions'', *Antipode*, 27(1): 49–70.

Mikkelson, B. (1995) *Methods for Development Work and Research*, London: Sage.

Milofsky, C. (2006) 'The Catalyst Process: What academics provide to practitioners', *Non-Profit Management & Leadership*, 16(4): 467–80.

Mitchell, D. (2002) 'Cultural landscapes: the dialectical landscape – recent landscape research in human geography', *Progress in Human Geography*, 26(3): 381–89.

Mitchell, D. and Staeheli, L. (2005) 'The complex politics of relevance in geography', *Annals of the Association American Geographers*, 95(2): 357–72.

Mohammad, R. (2001) ''Insiders' and/or 'Outsiders': Positionality, theory and praxis', in M. Limb and C. Dwyer (eds) *Qualitative Methodologies for Geographers: Issues and Debates,* Melanie Limb and Claire Dwyer, London: 101–20.

Mohan, G. (1999) 'Not so distant, not so strange: the personal and the political in participatory research', *Ethics, Place and Environment*, 2(1): 41–54.

—— (2001) 'Beyond participation: strategies for deeper empowerment', in B. Cooke and U. Kothari (eds) *Participation: The New Tyranny?*, London: Zed: 153–67.

—— (2007) Participatory development: from epistemological reversals to active citizenship, *Geography Compass*, 1(4). (in press)

Mohan, G. and Stokke, K. (2000) 'Participatory development and empowerment: the dangers of localism', *Third World Quarterly*, 21(2): 247–68.

Mohanty, C (2003) ''Under western eyes' revisited: feminist solidarity through anticapitalist struggles', *Signs: Journal of Women in Culture and Society*, 28(2): 499–535.

Morris, J. (1994) 'Gender and disability', in S. French (ed.) *On Equal Terms: Working with Disabled People*, London: Butterworth Heinemann: 207–20.

—— (ed.) (1996) *Encounters with Strangers: Feminism and Disability*, London: The Women's Press.

Moss, P. (ed.) (2002) *Feminist Geography in Practice: Research and Methods*, Oxford: Blackwell.

Moss, P. (forthcoming) 'Positioning a feminist supervisor in graduate supervision', *Journal of Geography in Higher Education.*
Mosse, D. (1994) 'Authority, gender and knowledge: theoretical reflections on the practice of participatory rural appraisal', *Development and Change*, 25(3): 497–526.
Mosse, D. (2000) 'People's Knowledge', Participation and Patronage: Operations and Representations in Rural Development', in B. Cooke and U. Kothari (eds) *Participation: The New Tyranny?*, London: Zed: 16–36.
Mountz, A., Miyares, I., Wright, R. and Bailey, A. (2003) 'Methodologically becoming: power, knowledge and team research', *Gender Place and Culture,* 10(1): 29–46.
Nairn, K., Higgins, J. and Sligo, J. (forthcoming) 'Youth researching youth: 'trading on' subcultural capital in peer research methodologies', *Teachers College Record.*
Nash, C. (1994) 'Remapping the body/land: new cartographies of identity, gender, and landscape in Ireland', in Rose, G. and Blunt A. (eds) *Writing Women And Space: Colonial And Postcolonial Geographies*, Guildford: The Guilford Press: 227–50.
—— (1996) 'Reclaiming vision: looking at landscape and the body', *Gender, Place, and Culture*, 32(2): 149–69.
—— (1997) 'Irish geographies: six contemporary artists, Nottingham', Exhibition, Djanogoly Art Gallery, University of Nottingham.
Nast, H. (1994) 'Opening remarks on "Women in the field"', *The Professional Geographer*, 46(1): 54–66.
Newman, M. (2006) *Teaching Defiance: Stories and Strategies for Activist Educators*, London: Wiley.
Nilan, P. (2002) "Dangerous fieldwork' re-examined: The question of researcher subject position', *Qualitative Research*, 2(3): 363–86.
Noddings, N. (1995) 'Caring', in V. Held (ed.) *Beyond Justice and Care: Essential Readings in Feminist Ethics*, Boulder, Colorado: Westview Press: 7–30.
O'Neill, M., Woods, P. and Webster, M. (2005) 'New arrivals: participatory action research, imagined communities and "visions" of social justice', *Social Justice*, 32(1): 75–88.
Oakley, A. (1981) 'Interviewing women: a contradiction in terms', in H. Roberts (ed.) *Doing Feminist Research*, London: Routledge: 30–61.
Offen, K. (2003) 'Narrating place and identity, or mapping Miskitu land claims in NE Nicaragua', *Human Organization*, 62(4): 382–92.
Oldenburg, R. (1999) *The Great Good Place*, New York: Marlowe and Company.
Oldfather, P. (1995) "Songs come back to most of them': students' experiences as researchers', *Theory into Practice,* 34(2): 131–37.
Oliver, K. (2001) *Witnessing: Beyond Recognition*, Minneapolis: University of Minnesota Press.
Oliver, M. (1992) 'Changing the social relations of research production', *Disability Handicap and Society*, 7(2): 101–14.
—— (1997) 'Emancipatory Research: Realistic goal or impossible dream?', in C. Barnes and G. Mercer (eds) *Doing Disability Research*, Leeds: The Disability Press: 15–31.
Ostrom E. (1990) *Governing the Commons: The Evolution of Institutions for Collective Action*. Cambridge: Cambridge University Press.
Pain, R. (2004) 'Social geography: participatory research', *Progress in Human Geography*, 28(5): 1–12.
—— (2006) 'Paranoid parenting? Rematerialising risk and fear for children', *Social and Cultural Geography,* 7(2): 221–43.
Pain, R. and Francis, P. (2003) 'Reflections on participatory research', *Area*, 35(1): 46–54.
Pain, R. and Kindon, S. (forthcoming) 'Participatory geographies', *Environment and Planning A*, 39. (in press)

Parfitt, T. (2004) 'The ambiguity of participation: a qualified defence of participatory development', *Third World Quarterly*, 25(3): 537–56.

Park, P., Brydon-Miller, M., Hall, B. and Jackson, T. (eds) (1993) *Voices of Change: Participatory Research in US and Canada*, London: Bergin and Harvey.

Parkes, M. and Panelli, R. (2001) 'Integrating catchment ecosystems and community health: the value of participatory action research', *Ecosystem Health*, 7(2): 85–106.

PLA Notes (1988–present) *Participation, Learning & Action Notes*: *Volume 1–49*, London: International Institute for Environment and Development. Available www.iied.org/NR/agbioliv/pla_notes/backissues.html (accessed 12 March 2007).

—— (2003) *Editorial Statement*, December, London: International Institute for Environment and Development. Available www.iied.org/NR/agbioliv/pla_notes/pla_backissues/48.html (accessed 31 May 2007).

Participatory Geographies Working Group (PyGyWG) of the Royal Geographical Society, UK. Available www.pygywg.org (accessed 20 June 2006).

Paterson, B.L., Gregory, D. and Thorne, S. (1999) 'A protocol for researcher safety', *Qualitative Health Research*, 9(2): 259–69.

Peet, R. (1969) 'A new left geography', *Antipode*, 1(1): 3–5.

Peluso, N. (1995) 'Whose woods are these? Counter-mapping forest territories in Kalimantan, Indonesia', *Antipode*, 27(4): 383–406.

People's Geographies project at Syracuse University (2007) Available www.peoplesgeographies.org (accessed 19 June 2006).

Phillips, O. (2004) 'The invisible presence of homosexuality: implications for HIV/AIDS and rights in Southern Africa', in Kalipeni, E., Craddock, S., Oppong, J.R. and Ghosh, J. (eds) *HIV/AIDS in Africa: Beyond Epidemiology*, Oxford: Blackwell Publishing: 155–66.

Pink, S. (2001) 'More visualising, more methodologies: on video, reflexivity and qualitative research', *The Sociological Review*, 49(2): 586–99.

—— (2007) *Doing Visual Ethnography*, 2nd edn, London: Sage.

Potts, K. and Brown, L. (2005) 'Becoming an Anti-Oppressive Researcher', in L. Brown and S. Strega (eds) *Research as Resistance: Critical, Indigenous, and anti-oppressive approaches,* Toronto: Canadian Scholars' Press/ Women's Press: 225–86.

Pratt, G. (1998) 'Comments on activism, in lost and found in the posts: addressing critical human geography', *Environment and Planning D: Society and Space,* 16(3): 264–5.

—— (2000) 'Participatory action research', in Johnston, R., Gregory, D., Pratt, G. and Watts, M. (eds) *The Dictionary of Human Geography*, 4th edn, Oxford: Blackwell: 574.

—— (2004) *Working Feminism*, Philadelphia: Temple University Press.

Pratt, G., in collaboration with the Philippine Women Centre (1999) 'Is this Canada? Domestic workers' experiences in Vancouver, BC', in J. Momsen (ed.) *Gender, Migration and Domestic Service,* London: Routledge: 23–42.

Pratt, G. in collaboration with the Ugnayan ng Kabataang Pilipino sa Canada/Filipino-Canadian Youth Alliance (Winter 2003/04) 'Between homes: displacement and belonging for second-generation Filipino-Canadian youths', *BC Studies*, 140: 41–66.

Pratt, G. in collaboration with the Philippine Women Centre (2005) 'From migrant to immigrant: domestic workers settle in Vancouver, Canada', in L. Nelson and J. Seager (eds) *Companion to Feminist Geography*, Oxford: Blackwell: 123–37.

Pratt, G. and Rosner, V. (2006) *The Global and the Intimate*, Guest-edited issue of *WSQ* (*Women's Studies Quarterly*), 34: 1–2.

Pretty, J., Guijt, I., Thompson, J. and Scoones, I. (eds) (1995) *Participatory Learning and Action: A Trainer's Guide*, London: International Institute for Environment and Development.

Pulido, L. (2003) 'The inner life of politics', *Ethics, place and environment*, 6(1): 46–52.
Punch, M. (1994), 'Politics and ethics in qualitative research', in N. Denzin and Y. Lincoln (eds) *Handbook of Qualitative Research*, Thousand Oaks: Sage: 83–97.
Quoss, B., Cooney, M. and Longhurst, T. (2000) 'Academics and advocates: using participatory action research to influence welfare policy', *The Journal of Consumer Affairs*, 34(1): 47–61.
Rahman, M. (1985) 'The theory and practice of participatory action research', in O. Fals-Borda (ed.) *The Challenge of Social Change*, London: Sage: 107–32.
Reardon, K. (1997) 'Participatory action research and real community-based planning in East St. Louis, Illinois', in P. Hyden, A. Figert, M. Shibley and D. Burrows (eds) *Building Community: Social Science in Action*, Thousand Oaks, CA: Pine Forge Press, 233–239.
Reason, P. (2001) 'Learning and change through action research', in J. Henny (ed.) *Creative Management*, London: Sage.
—— (2004) 'Action research and the single case: A response to Bjorn Gustavsen', *Concepts and Transformation*, 8(3): 281–94.
Reason, P. and Bradbury, H. (eds) (2006) *Handbook of Action Research*, London: Sage.
Richardson, L. (1993) 'Writing: a method of inquiry', in Y. Lincoln and N. Denzin (eds) *Handbook of Qualitative Research*, Thousand Oaks, CA: Sage Publishers: 516–29.
Rios-Moore, I., Arenas, E., Contreras, J., Jiang, N., Threatts, T, Allen, S. and Cahill, C. (2004) *Makes Me Mad: Stereotypes of Young Urban Womyn of Color*, New York: Center for Human Environments, Graduate School and University Center, City University of New York.
Robinson, J. (1994) 'White women researching/representing 'others': from antiapartheid to postcolonialism?', in G. Rose and A. Blunt (eds) *Writing Women and Space*, New York: The Guilford Press: 197–226.
Robson, C. (2002) *Real World Research*, Oxford: Blackwell.
Rohd, M. (1998) *Theatre for Community, Conflict & Dialogue: The Hope is Vital Training Manual*, Portsmouth, NH: Heinemann.
Rose, G. (1997) 'Situating knowledges: positionality, reflexivities and other tactics', *Progress In Human Geography*, 21(3): 305–20.
—— (2001) *Visual Methodologies*, London: Sage.
Routledge, P. (1996) 'The third space as critical engagement', *Antipode*, 28(4): 399–419.
—— (2002) 'Travelling East as Walter Kurtz: identity, performance, and collaboration in Goa, India', *Environment and Planning D: Society and Space*, 20(4): 477–98.
Rowlands, J. (1997) *Questioning Empowerment: Working with Women in Honduras*, Oxford: Oxfam Publications.
St. Martin, K. (2005) 'Mapping economic diversity in the first world: the case of fisheries', *Environment and Planning A*, 37(6): 959–79.
—— (2006) 'The impact of 'community' on fisheries management in the U.S. northeast', *Geoforum*, 37(2): 169–84.
Sampson, H. (2004) 'Navigating the waves: the usefulness of a pilot in qualitative research', *Qualitative Research*, 4(3): 383–402.
Sampson, H. and Thomas, M. (2003) 'Risk and responsibility', *Qualitative Research*, 3(2): 165–89.
Sanderson, E. and Kindon S. (2004) 'Progress in participatory development: opening up the possibilities of knowledge through progressive participation', *Progress in Development Studies*, 4(2): 114–26.
Sayer, A. (2000) *Realism and Social Science*, London: Sage.

Schaffer, K. and Smith, S. (2004) *Human Rights and Narrated Lives: The Ethics of Recognition*, Palgrave MacMillan, Basingstoke.

Schama, S. (1995) *Landscape and Memory*, London: Fontana Press.

Schensul, S., LoBianco, L. and Lombardo, C. (2004) 'Youth participatory research (youth PAR) in public schools: opportunities and challenges in an inner-city high school', *Practicing Anthropology*, 26(2): 10–14.

Selener, D. (1997) *Participatory Action Research and Social Change*, Ithaca, New York: Cornell Participatory Action Research Network, Cornell University.

Sen, S. and Nielsen, R. (1996) 'Fisheries co-management: a comparative analysis', *Marine Policy*, 20(5): 405–18.

Sense, A. (2006) 'Driving the bus from the rear passenger seat: control dilemmas in participative action research, *International Journal of Social Research Methodology*, 9(1): 1–13.

Servaes, J. (1996) 'Introduction', in J. Servaes, T. Jacobson and S. White (eds) *Participatory Communication for Social Change*, New Delhi: Sage: 13–27.

Sharp, J., Routledge, P., Philo, C. and Paddison, R. (eds) (2000) *Entanglements of Power: Geographies of Domination/Resistance*, London: Routledge.

Sheppard, E. (1995) 'GIS and society: towards a research agenda', *Cartography and Geographic Information Systems*, 22(2): 5–16.

Sieber, R. (2004) 'Towards a PPGIScience?', *Cartographica*, 38(3/4): 1–4.

da Silva, P. (2004) 'From common property to co-management: lessons from Brazil's first maritime extractive reserve', *Marine Policy*, 28(5): 419–28.

Silver-Pacuilla, H. and Associates from the Women in Literacy Project (2004) 'The meanings of literacy: a participatory action research project involving women with disabilities', *Women's Studies Quarterly*, 32(1/2): 43–58.

Sivananda, S. (2007) *Mahatma Ghandi (1869–1948)*. Available www.dlshq.org/saints/gandhi.htm (accessed 30 May 2007).

Slocum, R., Wichhart, L., Rocheleau, D. and Thomas-Slayter, B. (eds) (1995) *Power, Process and Participation: Tools for Change*, London: Intermediate Technology Ltd.

Smith, L. (1999) *Decolonizing Methodologies: Research and indigenous peoples,* New York: Zed.

Smith, R., Monaghan, M. and Broad, B. (2002) 'Involving young people as co-researchers: facing up to the methodological issues', *Qualitative Social Work*, 1(2): 191–207.

Solnit, R. (2004) *Hope in the Dark*, New York: Nation Books.

Spivak, G. (1988) 'Can the subaltern speak?', in C. Nelson and L. Grossberg (eds) *Marxism and the Interpretation of Culture*, Chicago: University of Illinois Press: 271–315.

Staeheli, L. and Lawson, V.A. (1994) 'A discussion of "Women in the Field": the politics of feminist fieldwork', *The Professional Geographer*, 46(1): 96–102.

Staeheli, L. and Mitchell, D. (2005) 'The complex politics of relevance in geography', *Annals of the Association of American Geographers*, 95(2): 357–72.

Standing, G. (1989) 'Global feminization through flexible labor', *World Development*, 17(7): 1077–95.

Stocks, A. (2003) 'Mapping dreams in Nicaragua's Bosawas reserve', *Human Organization*, 62(4): 344–56.

Stoeker, R. (1999) 'Are academics irrelevant? Roles for scholars in participatory research', *American Behavioral Scientist*, 42(5): 840–54.

Storper, M. and Scott, A. (1990) 'Work organisation and local labour markets in an era of flexible production', *International Labour Review*, 129(5): 573–91.

Strack, R., Magill, C. and McDonagh, K. (2004) 'Engaging youth through photovoice', *Health Promotion Practice,* 5(1): 49–58.

Stuart, S. and Bery, R. (1996) 'Powerful grass-roots women communicators: participatory video in Bangladesh', in J. Servaes, T. Jacobson and S.White (eds) *Participatory Communication for Social Change,* New Delhi: Sage: 197–212.

Suchet-Pearson, S. (2004) "I give you silence': situated engagement and dialogical research', paper presented at the 30th Congress of the International Geographical Union: IGC-UK 2004 Glasgow, August 2004.

Sunderlin, W., Angelsen, A., Belcher, B., Burgers, P., Nasi, R., Santoso, L., and Wunder, S. (2005) 'Livelihoods, forests, and conservation in developing countries: an overview', *World Development,* 33(9): 1383–402.

Taggart, R. (ed.) (1997) *Participatory Action Research: Contexts and Consequences,* Albany, New York: SUNY Press.

Taylor, R., Jason, L., Keys, C., Suarez-Balcazar, Y., Davis, M., Durlak, J. and Holtz Isenburg, D. (2004) 'Introduction: capturing theory and methodology in participatory research', in L. Jason, C. Keys, Y. Suarez-Balcazar, R.Taylor and M. Davis (eds), with J. Durlak and D. Holtz Isenburg, *Participatory Community Research: Theories and Methods in Action,* Washington, DC: American Psychological Association: 3–14.

Thomas, C. (1999) *Female Forms: Experiencing and Understanding Disability,* Buckingham: Open University Press.

Thornton, S. (1995) *Club Cultures: Music, Media and Subcultural Capital,* Cambridge: Blackwell.

Thrift, N. (1997) 'The still point: resistance, expressive embodiment and dance', in S. Pile and M. Keith (eds) *Geographies of Resistance,* London: Routledge: 124–51.

—— (2007) *Non-representational Theory: Space, Politics, Affect.* London: Routledge.

Torre, M.E, Fine, M., Boudin, K., Bowen, I., Clark, J., Hylton, D., Martinez, M., Roberts, R., Rivera, M., Smart, P. and Upegui, D. (2001) 'A space for co-constructing counter stories under surveillance', *International Journal of Critical Psychology,* 4: 149–66.

Townsend, J., with Arrevillaga, U., Bain, J., Cancino, S., Frenk, S., Pacheo, S. and Crez, E. (1995) *Women's Voices from the Rainforest,* London: Routledge.

Trapese (2007) *DIY: A handbook for changing our world,* London: Pluto.

US CENSUS BUREAU (2000) Community Resource Data. Available www.census.gov/main/www/cen2000.html (accessed 5 June 2007).

Van Blerk, L. and Ansell, N. (2007) 'Participatory feedback and dissemination *with* and *for* children: reflections from research with young migrants in Southern Africa', *Children's Geographies,* 5(3) 313–24.)

Van Vlaenderen, H. (1999) 'We live on prayers: the use of video in community development', *PLA Notes* 35: 3–6. Available www.poptel.org.uk/iied/sarl/pla_notes/pla_backissues/documents/plan_03501.pdf (accessed 29 May 2007).

Vernon, A. (1997) 'Reflexivity: the dilemmas of researching from the inside', in C. Barnes and G. Mercer (eds) *Doing Disability Research,* Leeds: The Disability Press: 158–76.

Wadsworth, Y. (1997) *Report on the Cartagena AR/PAR Congress,* Melbourne: The Action Research Issues Association.

—— (1998) 'What is participatory action research?', *Action Research International,* Paper 2. Available www.scu.edu.au/schools/gcm/ar/ari/p-ywadsworth98.html (accessed 25 March 2007).

—— (2000) 'The mirror, the magnifying glass, the compass and the map: facilitating participatory action research', in P. Reason and H. Bradbury (eds) *Handbook of Action Research,* London: Sage: 420–32.

—— (2006) 'The Mirror, the Magnifying Glass, the Compass and the Map: Facilitating Participatory Action Research', in P. Reason and H. Bradbury (eds) *Handbook of Action Research*, London: Sage: 420–32.

Wang, C. (1999) 'Photovoice: A participatory action research strategy applied to women's health', *Journal of Women's Health,* 8(2): 185–92.

—— (2003) 'Using photovoice as a participatory assessment and issue selection tool: a case study with the homeless in Ann Arbor', in M. Minkler and N. Wallerstein (eds) *Community-based participatory research for health*, San Francisco: Jossey-Bass: 179–96.

Wang, C. and Burris, M. (1997) 'Photovoice: concept, methodology, and use for participatory needs assessment', *Health Education and Behavior,* 24(3): 369–87.

Wang, C., Morrel-Samuels, S., Hutchison, P., Bell, L. and Pestronk, R. (2004) 'Flint photovoice: community building among youths, adults and policymakers', *American Journal of Public Health*, 94(6): 911–13.

Wates, M. and Jade, R. (eds) (1999) *Bigger Than the Sky: Disabled Women on Parenting*, London: The Women's Press.

Whatmore, S. (2006) 'Materialist returns: practising cultural geography in and for a more-than-human world', *Cultural Geographies,* 13(4): 600–9.

White, G. (1972) 'Geography and public policy', *The Professional Geographer* 24(May): 101–4.

WHO (2006) 'Epidemiological fact sheets on HIV/AIDS and sexually transmitted infections: Zimbabwe, 2006 Update', *World Health Organisation*, Geneva. Available www.who.int/GlobalAtlas/predefinedReports/EFS2006/EFS_PDFs/EFS2006_ZW.pdf (accessed 26 May 2007).

Whyte, W. (ed.) (1991) *Participatory Action Research*, Newbury Park: Sage.

Williams, G. (2004a) 'Evaluating participatory development: tyranny, power and (re)politicisation', *Third World Studies Quarterly,* 25(3): 557–78.

—— (2004b) 'Towards a repoliticization of participatory development: political capabilities and spaces of empowerment', in S. Hickey and G. Mohan (eds) *Participation: From Tyranny to Transformation? Exploring New Approaches to Participation in Development*, London: Zed: 92–108.

Willson, K. (2006) *Low Income Women Speak Out through Photovoice Projects in Winnipeg and Saskatoon.* Available www.pwhce.ca/program_poverty_photovoice.htm> (accessed 14 November 2006).

Wilshusen, P., Brechin, S., Fortwangler, C. and West, P. (2002) 'Reinventing a square wheel: critique of a resurgent "protection paradigm" in international biodiversity conservation', *Society & Natural Resources*, 15(1): 17–40.

# Index

Lightning Source UK Ltd.
Milton Keynes UK
UKHW01f0809110718
325542UK00013B/72/P

9 780415 599764